Life Cycle Engineering
of Plastics

Technology, Economy and the Environment

Life Cycle Engineering of Plastics

Technology, Economy and the Environment

Lars Lundquist
Yves Leterrier
Paul Sunderland
Jan-Anders E. Månson

*École Polytechnique Fédérale
de Lausanne, Switzerland*

2000
ELSEVIER
Amsterdam - Lausanne - New York - Oxford - Shannon - Singapore - Tokyo

ELSEVIER SCIENCE Ltd
The Boulevard, Langford Lane
Kidlington, Oxford OX5 1GB, UK

First edition 2000

Library of Congress Cataloging-in-Publication Data
Life cycle engineering of plastics: technology, economy and the environment / L. Lundquist ...[et al.]
        p. cm.
    Includes index.
    ISBN 0-08-043886-5 (hardcover)
    1. Plastics. 2. Product life cycle. I. Lundquist, L.

TP1122.L54 200
668.4--dc21   00-049486

British Library Cataloguing in Publication Data
Life cycle engineering of plastics: technology, economy and the environment
1. Plastics 2. Product life cycle
I. Lundquist, L.
668.4

ISBN 0-08-043886-5

Transferred to digital printing 2006
Printed and bound by Antony Rowe Ltd, Eastbourne

PREFACE

Technology as manifest in everyday products will continue to play a critical role in providing for human satisfaction even as societies become more and more aware of deleterious impacts on supporting ecosystems. Achieving a sustainable way of living requires those who design everyday artefacts to build-in features that protect the global life support system. For many today this need is clear, but the realization of sustainable products is just now emerging. The authors of this text offer a clear and comprehensive vision of life cycle engineering which provides a framework for design that accounts for impacts over the entire product life cycle.

In this book, Lundquist, Leterrier, Sunderland and Månson add much to the already evolving field of Design for Environment. But this book goes far beyond most works on this subject by surrounding the central notions of life cycle assessment with a scientific body of knowledge and with a more practical slant reflecting the reality of the organizations in which product development occurs. Through a focus on plastic products, the authors show the importance of making ties between basic technical knowledge and the process of life cycle engineering. Their approach offers a practical, deliberate way to make ecologically and economically sensible decisions about product reuse and recycling and other critical dimensions of product life behavior. They demonstrate a positive approach to designing products that fits into a sustainable economy through down-to-earth cases. While the book focuses on the life cycle engineering of plastics, it is only a short step to other materials and products.

Beyond contributing to the technology of life cycle engineering, this text adds to the growing body of knowledge that argues for an fundamentally new way of thinking about economic and social activity--a new paradigm for sustainable social and industrial problem solving. Industrial ecology is such a new system for thinking about and implementing sustainability that draws its core set of ideas from the ecological world. Industrial ecology brings to the surface the idea of interdependence among members of a community-- natural or economic, and notes the material cycles that are central to a stable ecosystem. The life cycle engineering framework, coupled with sound scientific knowledge of materials behavior as articulated in this book, makes a giant step towards bringing the model of industrial ecology into everyday practice.

JOHN R. EHRENFELD
Director, MIT Technology, Business and Environment Program
Center for Technology, Policy, and Industrial Development
Massachusetts Institute of Technology

PREAMBLE

There is widespread agreement that something has to be done to ensure an acceptable environment for future generations. There is nevertheless much debate as to how to go about it and under what conditions. It is an issue with wide-ranging implications in society, not the least of which is industrial activity. Whether a villain for some or a forward-looking, co-operative partner for others, the role of industry in initiating and implementing environmental change is crucial.

This book aims to present methods for dealing with environmental issues in the plastics industry. The concept of Life Cycle Engineering (LCE) can help to better understand the opportunities and challenges facing this industrial sector as well as the limits to possible forms of action. The industry can act towards improving the environment while at the same time maintaining a competitive edge.

Following an introduction to LCE in Chapter 1, the scientific basis for LCE is presented in Chapter 2. It is well-known that the properties of plastics change over time. A thorough understanding of the mechanisms behind these changes is necessary for improving the environmental performance of plastics and polymer composites throughout their life cycle.

Waste management and recycling alternatives are presented and discussed in Chapter 3. The poor reputation of used plastics is gradually dissipating, as methods to recover and recycle post-consumer waste plastics improve and reliable high-quality supplies of reusable material can be guaranteed.

The concept of Life Cycle Assessment (LCA) is described in Chapter 4. As an analytical instrument for generating information on resource consumption and environmental impact, LCA is increasingly finding acceptance for the evaluation of materials, processes and products in a wide range of situations.

Chapter 5 covers in detail the role of LCE in product and process development. As early an involvement as possible of environmental considerations in the development process is critical for minimising the overall environmental impact of the production, use, revaluation and disposal of products and processes.

Such improvements are not only determined by what is done but also, crucially, by how it is done and by whom. Thus, Chapter 6 describes the organisational aspects of Life Cycle Engineering. Close collaboration between a corporation's technical, managerial and financial functions is a prerequisite for the successful application of LCE practices.

The final Chapter 7 presents several case studies which demonstrate approaches for implementing LCE. The development and production of plastics products in several sectors of industry are outlined and dissensed in the light of the material covered in the previous six chapters.

It is our hope that this book will provide a practical and readable guide to environmental strategies for the plastics industry. In trying to appeal to readers with a broad range of backgrounds, we hope to encourage the collaboration between technologists, managers and financiers that is so important for the success of Life Cycle Engineering practices.

ACKNOWLEDGEMENTS

The authors would like to thank the companies having contributed with case studies:

Pharmacia & Upjohn, Stockholm, Sweden:

Bengt Mattson

Sven Göthe

Gösta Larsson

Volvo Car Corporation, Göteborg, Sweden:

Ulla-Britt Fräjdin Hellkvist

Ulf Jansson-Liljeroth

Carl-Otto Nevén

Inge Horkeby

AB Konstruktions-Bakelit, Örkeljunga, Sweden:

Anders Månson

Kent Eriksson

Anders Edsfeldt

DuPont de Nemours International S.A., Geneva, Switzerland:

Philip Boydell

Victor Williams

Our gratitude to our colleagues P.-A. Eriksson, Yves Wyser, Louis Boogh, Ronny Törnqvist, and Gustav Jannerfeldt who have provided technical discussions, information and collaboration in the areas covered in this book. Many thanks are also due to Prof. Leif Carlsson of the Florida Atlantic University for devoting some of his time reading through parts of the manuscript criticising the content, the form and the presentation.

We would like to extend our gratitude to George Haour, Francisco Szekely, and Ulrich Steger at the International Institute for Management Development, Lausanne, Switzerland for helpful exchanges and comments.

We are also indebted to Olivier Jolliet at L'Institut de l'Aménagement de la Terre et des Eaux, Département de Génie Rural, Ecole Polytechique Fédérale de Lausanne, Switzerland for instructive discussions and exchanges within the field of Life Cycle Assessment.

CONTENTS

PREFACE V

PREAMBLE VI

ACKNOWLEDGEMENTS IX

1 INTRODUCTION TO LIFE CYCLE ENGINEERING 1

    1.1 WHAT IS LIFE CYCLE ENGINEERING? 3

    1.2 THE GROWTH OF PLASTICS 4

    1.3 THE LIFE CYCLE ENGINEERING OF PLASTICS 7

    1.4 WHO SHOULD BE INVOLVED? 9

2 THE POLYMER LIFE CYCLE 13

    2.1 WHAT IS A POLYMER? 13

    2.2 WHAT ARE PLASTICS? 18

    2.3 DURABILITY AND RELIABILITY OF PLASTIC PRODUCTS 21

    2.4 DEGRADATION AND AGEING OF POLYMERS 23

        2.4.1 Process induced degradation 24

        2.4.2 Service-induced degradation 26

        2.4.3 Physical ageing 27

        2.4.4 Viscoelastic effects 29

    2.5 LIFE-TIME PREDICTION 30

        2.5.1 Long-term prediction of thermo-oxidative degradation 30

        2.5.2 Superpositions and shift factors 32

        2.5.3 Prediction of fatigue failure 34

        2.5.4 Material know-how and life cycle engineering 36

3 PLASTICS RECOVERY AND RECYCLING 39

    3.1 THE VITAL RECYCLING CHAIN 39

    3.2 COLLECTION AND SORTING FOR RECYCLING 40

        3.2.1 Identification and sorting systems 41

        3.2.2 Economics of collection and sorting 45

        3.2.3 Conclusions 47

    3.3 WASTE MANAGEMENT ROUTES 47

        3.3.1 Mechanical recycling 48

        3.3.2 Revitalisation 52

3.3.3 Feedstock recycling 57

3.3.4 Energy recovery 60

3.3.5 Environmentally-degradable polymers 63

3.4 APPLICATIONS FOR RECYCLED PLASTICS 65

3.4.1 Factors affecting market acceptance 65

3.4.2 Emerging markets for recycled plastics 68

3.4.3 Responsibility in the recycling chain 69

4 LIFE CYCLE ASSESSMENT **77**

4.1 INTRODUCTION 77

4.2 GOAL DEFINITION AND SCOPING.. 80

4.2.1 Purpose 80

4.2.2 Scope.. 81

4.2.3 Determining the functional unit...................................... 82

4.2.4 Data quality assessment 84

4.3 INVENTORY ANALYSIS 85

4.3.1 Generating the process tree... 85

4.3.2 Entering the process data 87

4.3.3 Applying the allocation rules 88

4.3.4 Creating the inventory table 90

4.4 IMPACT ASSESSMENT 91

4.4.1 Classification 91

4.4.2 Characterisation... 92

4.4.3 Evaluation 94

4.5 IMPROVEMENT ANALYSIS 95

4.6 CONCLUSIONS AND POINTERS 96

5 LIFE CYCLE ENGINEERING IN PRODUCT DEVELOPMENT **101**

5.1 RESOURCE OPTIMISATION 101

5.2 THE PRODUCT DEVELOPMENT PROCESS 103

5.2.1 Integrated product development 104

5.2.2 Computer-based "green" design tools 105

5.3 REDUCTION OF MATERIAL INTENSITY 106

5.3.1 Material selection 106

5.3.2 Weight reduction... 108

5.3.3 Material reduction through design.................................... 111

5.4 PRODUCT LIFE EXTENSION 115

5.4.1 Modularity and maintainability 116

5.4.2 Repair, remanufacturing and reuse 117

5.5 MATERIAL LIFE EXTENSION 119

   5.5.1 Material optimisation.. 119

   5.5.2 Design for recycling 121

5.6 PERSPECTIVES.. 124

**6 ORGANISATIONAL ASPECTS OF LIFE CYCLE
ENGINEERING** **129**

6.1 THE IMPORTANCE OF NETWORKING 129

6.2 THE IMPORTANCE OF COMMUNICATION 132

6.3 ENVIRONMENTAL MANAGEMENT.................................... 134

   6.3.1 The ICC Business Charter for Sustainable Development 134

   6.3.2 BS 7750 137

   6.3.3 The EEC Environmental Management and Audit Scheme
   (EMAS)... 139

   6.3.4 ISO 14000 142

   6.3.5 Core elements in EMS and environmental reporting 143

   6.3.6 Cost and utility of environmental management.......................... 147

6.4 EMS IN SMALL AND MEDIUM SIZED COMPANIES 149

6.5 TOWARDS GREENER MANAGEMENT. A SUMMARY...................... 151

7 CASE STUDIES **155**

7.1 PACKAGING RESTRUCTURING.. 155

   7.1.1 Pharmacia & Upjohn.. 155

   7.1.2 Packaging development 156

   7.1.3 Packaging recycling 158

   7.1.4 Attitudes of hospital personnel.. 162

   7.1.5 The vital recycling chain 163

7.2 INDUSTRIAL NETWORKING FOR COMPETITIVENESS 164

   7.2.1 Plastics in the automotive industry 164

   7.2.2 Volvo Car Corporation 168

   7.2.3 AB Konstruktions-Bakelit 177

   7.2.4 DuPont de Nemours International S.A. 182

   7.2.5 Future automotive recycling networks 191

8 AFTERWORD **195**

ACRONYMS **199**

INDEX **203**

1

INTRODUCTION TO LIFE CYCLE ENGINEERING

It is no longer physically possible to externalise the environmental costs and damages outside the production process and allow them to be borne by nature and future generations: industrial processes and products must be redesigned to internalise such costs and damages within the production and consumption processes.

"The Limits to competition", The Group of Lisbon [1].

The preservation of non-renewable natural resources is an issue which has assumed great importance in modern society. Our "throw-away" culture, characterised by a linear treatment of assets which begins with natural resources and ends in waste, is perceived as being guilty of wasteful squandering of precious natural resources.

There is strong pressure to take remedial action. Its importance and urgency was expressed as early as 1972, when The Club of Rome published "The Limits to Growth" [2]. It is now widely recognised that continued economic development should be accompanied by more appropriate use of natural resources. Just how the economy should be organised to achieve such balanced growth is at present open to a great deal of debate [3].

Industry is under great pressure to improve its practices. The polymer industry is particularly under fire, no doubt due to the short lifespan of many plastics-based consumer products, the high visibility of polymers in municipal solid waste and the rapid increase of plastics consumption.

In the past the strategy of organisations facing up to environment-related issues has largely involved taking action of a "fire-fighting" nature, resolving symptoms rather than causes. Finding themselves forced into positions of weakness, companies have tended to see the financial and organisational efforts required as pure costs. This is changing: *the more forward-looking environmental performers are now considering the environment as a creator of business opportunity rather than as a constraint* [4]. Industrial acceptance of the importance of environmental issues is clear in the creation of organisations such as the World Business Council for Sustainable Development (WBCSD) [5]. The role of this coalition of more than 120 international companies is proclaimed to be to provide business leadership as a catalyst towards sustainable development and to promote the attainment of eco-efficiency through high standards of environmental and resource management in business.

Environmental campaigners are adopting a more subtle approach towards industry. Pressure groups such as Greenpeace grew through aggressive opposition to industry, to bring environmental problems to the attention of the public and decision-makers. It is acknowledged that in some cases this impeded constructive change, and there is now a tendency to work together with industry to promote environmentally-sound practices and to solve problems by collaboration. An example result of this policy is the CFC-free refrigerator, pioneered by Greenpeace and manufactured by several leading white goods manufacturers [6, 7].

Sustainable development has been widely accepted as a viable concept for ensuring long-term survival of mankind and of other species on the planet under acceptable conditions [8-18]. Although there is widespread acceptance of the concept there are still large discrepancies concerning its implications. The term was first adopted by the Brundtland Commission, responsible for establishing a global plan of action (Agenda 21) after the Rio environmental summit of 1992 [19]. It emphasises the relationship between present and future human needs and nature's capacity to meet those needs [20]:

Sustainable development is development that meets the needs of the present without compromising the ability of future generations to meet their own needs. It contains within it two key concepts:

- *the concept of "needs", in particular the essential needs of the world's poor, to which overriding priority should be given;*

- *the idea of limitations imposed by the state of technology and social organisations on the environment's ability to meet present and future needs.*

Sustainable growth must therefore encompass environmental, social, and economic factors and maintain a balance between them. This implies consideration of a wide range of factors in determining solutions to environmental problems. Although the public eye is often focused only on waste management and recycling, rigorous and correct technical solutions require that the overall, much broader, picture be studied.

1.1 WHAT IS LIFE CYCLE ENGINEERING?

Every year 2,800,000,000 tonnes of all categories of waste are produced in Western Europe. Assuming an average density of 2 kg/l, this waste would cover a surface area equivalent to that of Switzerland to a depth of three centimetres.

One of the preconditions for further sustainable economic growth is that the implementation of new technology should improve the efficiency of resource consumption and reduce waste output. To be most effective, improvements should occur at all the stages of the life cycle of a product, from raw material production, part manufacture and use to recycling or disposal.

LCE provides a methodology of how to design, manufacture, use, maintain and recover materials and products with the aim of optimising resource use and minimising environmental impact (Figure 1.1).

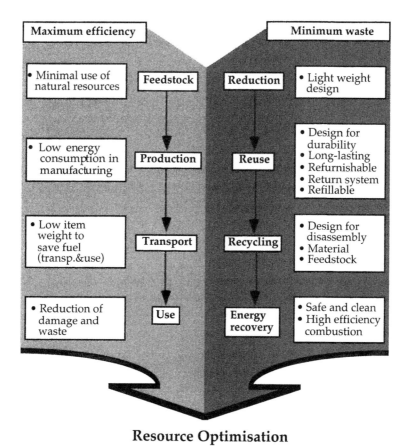

Resource Optimisation

Figure 1.1   Life cycle engineering practices (after APME [21]).

Not only technical but also economic, legislative, emotional and political factors influencing the life cycle must be taken into account. LCE places industrial activities at the intersection of politics, the environment, the economy and technology (Figure 1.2). Industry is well-positioned to lead environmental change, as it possesses thorough knowledge of the possibilities and limitations of the application of technology within the market economy. Economical and environmental progress should be possible together if LCE practices are implemented.

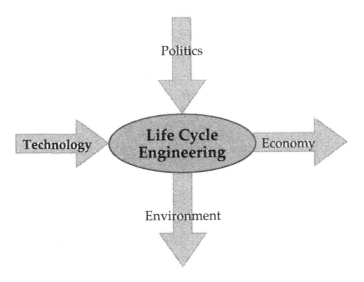

Figure 1.2   Life cycle engineering is a technological process within an economical, environmental and political framework.

1.2 THE GROWTH OF PLASTICS

The plastics sector has undergone rapid development in recent times. From 1.5 million tonnes in 1945, annual production had grown to approximately 80 million tonnes in 1989 [22]. This can be compared to the production of a mature material such as pig iron, which has grown approximately four-fold over the same time period.

Nowadays plastics are used within most sectors of industry, the predominant sectors in terms of tonnage being packaging and building / construction (Figure 1.3).

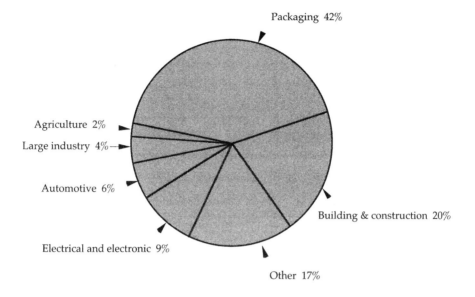

Figure 1.3   Plastics consumption by industry sector in Western Europe [23].

The plastics industry promotes the use of plastics in Europe through the Association of Plastics Manufacturers in Europe (APME) [23]. The advantages of plastics over traditional materials are often cited as being their [24]:

- versatility,
- low weight,
- transparency,
- safety and hygiene,
- cost-effectiveness,
- durability.

Their versatility has lead to the development of the approximately 20,000 different types and grades of plastics presently on the market. Based on the major types of polymers, as shown in Table 1.1, modification by copolymerisation, blending, alloying or mixing with additives means that plastics can be tailored to suit the requirements of individual applications. They can also be combined in complex assemblies to provide multiple functionality. Examples of this are multilayer packaging and automotive interior trimmings. The latter are triple layer structures composed of a rigid backing material, an intermediate tactile-enhancing layer and a surface material such as a textile or PVC for surface appearance.

Table 1.1   Major categories of plastics on the market.

Thermoplastics	
Engineering resins	**Commodity resins**
Nylons (PA)	*Polyolefines*
Polyacetals	Polyethylene (PE)
Polysulfones	low density (LDPE)
Polysulfides	linear low density (LLDPE)
Polyetheretherketone (PEEK)	high density (HDPE)
Polyphenyleneoxide (PPO)	Polypropylene (PP)
Polyimides (PI)	
Polyvinyl chloride (PVC)	Polystyrene (PS)
Polyurethanes (PU)	
Polyetheneterephtalate (PET)	
Polycarbonate (PC)	

Thermosets (Crosslinked polymers)	
Thermosets	**Rubbers**
Phenol-formaldehyde (PF)	Natural rubber (NR)
Urea-formaldehyde (UF)	Butyl rubber (IIR)
Melamine-formaldehyde (MF)	Styrene-butadiene rubber (SBR)
Epoxy	Acrylonitrile-butadiene rubber (NBR)
Unsaturated polyester (UP)	Neoprene (CR)
	Polyisoprene (IR)
	Polyurethane elastomers (PU)

The use of plastics within the packaging and automotive sectors has allowed considerable weight- and energy savings without a manufacturing cost penalty.

The industry is nonetheless a target of choice for environmental criticism, as it is claimed that plastics consume non-replenishable natural resources in the form of petroleum, the base feedstock for the synthesis of plastics. Given that the annual petroleum consumption of the plastics-producing industry corresponds to approximately 4% of the total petroleum production (Figure 1.4) and that the remaining part is used as a combustible for heating and transport, the industry feels unfairly exposed to criticism. Nevertheless, it has recognised the importance of every actor in society carrying is part of the burden.

The qualities of plastics can also be disadvantageous. Their high durability, often used as a sales argument, becomes a problem in applications where the service life is short, such as packaging. Plastics packaging waste makes up a considerable and very visible share of municipal solid waste (MSW) and litter.

Rising concern for resource minimisation and waste management is dictating new priorities for polymer-based materials and products. It is becoming increasingly important to take environmental issues into consideration when designing such a product, rather than only economic and performance criteria.

What will happen to the product when it is replaced and who is responsible for it? Such considerations change the way materials are selected and assembled.

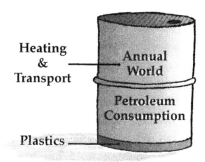

Figure 1.4   Annual world petroleum consumption including petroleum consumed in the manufacture of plastics.

It is no longer acceptable to design complex parts and assemblies without considering the environmental impacts of their production, use, and disposal. If a part is to be recycled, for example, it will not be possible to retain material quality without separating incompatible materials. Or, if materials have been assembled in a complex manner, it may be too costly to separate and recycle to the required level of quality.

The complexity of environmental issues related to plastics and the risk of facing harmful public and legislative pressure have encouraged many major chemical companies and plastics processors and users to be proactive by adopting a life cycle-based strategy for their activities. These companies are now at the forefront of the environmental debate between industry, governments, non-governmental organisations and universities.

1.3 THE LIFE CYCLE ENGINEERING OF PLASTICS

The life cycle of a product made from a polymer or from a polymer-based composite is composed of a series of distinct steps (Figure 1.5). The synthesised polymer possesses the highest performance value per definition. This value is a measure of the properties and durability the polymer can offer and hence of its usefulness as a material, rather than of its monetary value. The material undergoes several operations which modify its form and function. During each step a thermal, mechanical or chemical load is imposed on the material, which tends to reduce its intrinsic performance value.

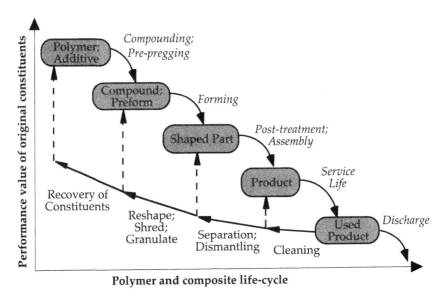

Figure 1.5   Performance value of polymers and composites throughout their life cycle.

When the material is incorporated into a part or an assembly it can become far more difficult to recuperate for subsequent reuse. A polymer painted with an incompatible coating, for example, may lose all its performance value at the painting stage since the chances of separating it from the paint may practically be nil.

Often a used item of packaging has a performance value of zero since it is likely to be discharged into a landfill. This is not the only possible scenario, however, nor is it likely to remain a viable one for much longer. Alternatively, several passes through all or parts of the life cycle loop may be envisaged. This can be achieved by chemical recycling, which converts the material into monomers, its original constituents, by incineration with energy recovery or by mechanical recycling. In the two first scenarios, however, the material value is lost. It is only mechanical recycling that conserves part of the material performance value, provided that it has not dropped too low during the preceding life cycle stages.

By monitoring the value of the polymer throughout its life cycle, from the "cradle" to the "grave", it is possible to guide improvement, the aim being to maintain the performance value of the constituent materials at as high a level as possible throughout the life cycle as illustrated in Figure 1.6 by the dark-shaded boxes. This is the goal of Life Cycle Engineering.

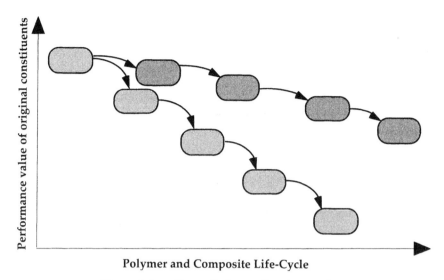

Polymer and Composite Life-Cycle

Figure 1.6   Controlled cascading in recycling.

1.4 WHO SHOULD BE INVOLVED?

Many diverse players are involved in the environmental debate in society as shown in Figure 1.7: governmental institutions, consumer organisations, lobbyists, pressure groups, industry, universities and financial institutions, to name several. Their goals can, of course, differ significantly.

Figure 1.7   Environmental players in society.

Within industry the principal actors representing the technical, managerial and financial functions must collaborate actively if Life Cycle Engineering practices are to be implemented. Change cannot be undertaken by a single group alone, as its effects go far beyond any one function's activities. This collaboration should occur as early as possible during product development.

Each employee of a corporation is also, of course, a member of society and as such he or she must reconcile personal views on environmental change with LCE practice. The commitment of management within a corporation to Life Cycle Engineering is a necessary, but not sufficient, condition for ensuring its acceptance at all levels of the company. Environmental change within the corporation cannot occur if it proves to be out of phase with change in society.

Increased public attention to environmental issues has brought about rapid change in legislation and in consumer patterns, with immediate consequences for industrial activities. The response of organisations can take one of three possible forms:

- remain passive, and not respond to environmental legislation or to consumer pressure;
- comply to legislation;
- adopt a pro-active position by co-operating with authorities and with environmental and consumer organisations to anticipate legislation and shifting consumer trends.

Steadily increasing fines for non-compliance with laws and rocketing charges for dumping waste make the first alternative highly risky. The difficulty in following the second option was expressed by G.V. Cox of the Chemical Manufacturers' Association [25]:

"If someone would just tell us what to do once, and let us do it, we would be happy. But they keep changing the goals as we're going through the process, which makes compliance very costly and not necessarily very productive".

The solution is to be pro-active. Staying at the forefront of the debate may be costly but it can lead to significant benefits. Leading environmentally-active companies have implemented systematised practices to ensure compliance to current and future environmental legislation [26-28]. Among these practices are environmental management, Life Cycle Assessment (LCA) and Life Cycle Engineering (LCE). These aid in identifying and establishing opportunities for environmental improvement, technological and organisational structures for bringing change about, and in communicating progress to environmental stakeholders.

REFERENCES

1. *The limits to Competition*, The Group of Lisbon Ed., The MIT Press: Cambridge, USA (1995).

2. D. L. Meadows, D. H. Meadows, J. Randers, and W. B. Behrens, *The Limits to Growth: Report to the Club of Rome's Project on the Predicament of Mankind*, Potomac Associates, Universe Books: New York, USA (1972).

3. O. Giarini and W. R. Stahel, *The Limits to Certainty: Facing Risks in the New Service Economy*, 2nd ed., Kluwer Academic Publisher: Amsterdam, The Netherlands (1993). Also in French by PPUR, Lausanne (1990).

4. S. Schmidheiny, *Changing Course: A Global Business Perspective on Development and the Environment*, MIT Press: London (1992).

5. *Two Leading Business Organisations to Merge*, WICE, 40, Cours Albert 1er, 75008 Paris, France, Press release (1994).

6. M. Abrahamsson, *Electrolux Adopts "greenfreeze" technology*, Greenpeace Business, 21, p. 6 (1994).

7. C. Millais, *Electrolux Adopts "greenfreeze" technology*, Greenpeace Business, 18, p. 3 (1994).

8. *Environmental Reporting: A Manager's Guide*, World Industry Council for the Environment (WICE), Paris, France (1994).

9. *Company Environmental Reporting: A Measure of the Progress of Business and Industry Towards Sustainable Development*, UNEP/SustainAbility, Ltd, Technical report Nr. 24 (1993).

10. *Plastics Recovery in Perspective: Plastics Consumption and Recovery in Western Europe 1993*, Association of Plastics Manufacturers in Europe (APME), Brussels, Belgium (1995).

11. *WICE: World Industry Council for The Environment: Presentation Leaflet*, World Industry Council for the Environment (WICE): Paris, France (1994).

12. *Design for Environment*, World Industry Council for the Environment (WICE), Paris, France (1994).

13. *L'Etat de l'Environnement (1972-1992). Sauvons la Planète: Défis et Espoirs*, Programme des Nations-Unies pour l'Environment (PNUE), Nairobi, Kenya, UNEP/GCSS.III/2 (1992).

14. E. El-Hinnawi, *PNUE: Deux Décennies de Réalisations et de Défis*, Programme des Nations-Unies pour l'Environment (PNUE), Nairobi, Kenya (1992).

15. *Plastics Packaging: Friend or Enemy*, Association for Plastics Manufacturers in Europe (APME), Brussels, Belgium.

16. J. Nüesch, *Preface of the 3rd International Seminar on Life Cycle Engineering* in proceedings of *ECO-Performance*, Zurich, Switzerland, Verlag Industrielle Organisation, Zurich, Switzerland, pp. 3-4 (1996).

17. L. Bern, *Uthålligt Ledarskap*, Ekerlids Förlag: Kristianstad, Sweden (1993).

18.   K.-H. Robèrt, *Det Nödvändiga Steget*, 3rd ed., Affärsförlaget Mediautveckling: Kristianstad, Sweden (1993).

19.   *Agenda 21: Programme of Action for Sustainable Development* in proceedings of *United Nations Conference on Environment and Development (UNCED)*, Rio de Janeiro, Brazil, United Nations, Departement of Public Information, New York (1993).

20.   G. H. Brundtland, *How to Secure our Common Future*, Scientific American, September, p. 134 (1989).

21.   *Plastics Recovery in Perspective: Plastics Consumption and Recovery in Western Europe 1992*, Association of Plastics Manufacturers in Europe (APME), Brussels, Belgium (1994).

22.   C. J. Williamson, *Victorian Plastics-Foundations of an Industry* in proceedings of *Symposium on the History of Synthetic Materials in conjunction with the Annual Chemical Congress of the Royal Society of Chemisty*, Southampton, UK, The Royal Society of Chemistry, pp. 1-9 (1993).

23.   *Plastics: A Material of Choice for the 21th Century: Insight into Plastics Consumption and Recovery in Western Europe 1997*, Association of Plastics Manufacturers in Europe (APME), Brussels, Belgium (1999).

24.   *Plastics: A Vital Ingredient for the Food Industry*, Alliance of Plastic Packaging for Food (APPF). Association for Plastics Manufacturers in Europe (APME), Brussels, Belgium.

25.   D. R. Cannon and L. A. Rich, *Hazardous Waste Management - New Rules are Changing the Game*, Chemical Week, **139**, 8, pp. 26-64 (1986).

26.   DuPont, *Environmentalisme d'entreprise: Bilan Europe 1993*, DuPont de Nemours International S.A. Environmental Affairs-Europe Le Grand-Saconnex, Switzerland, 1 (1993).

27.   *Environmental Performance Report*, Rank Xerox, November (1995).

28.   *Environmental Report*, National Westminster Bank Plc. (1994).

2

THE POLYMER LIFE CYCLE

In this chapter the basic features of polymers and their additives are introduced. The main ageing and degradation phenomena, which relate to the reliability and durability of polymers, from manufacture and service, to final disposal or recovery, are reviewed. The prediction of the long-term performance of polymers is discussed.

2.1 WHAT IS A POLYMER?

Plastic products are based on polymers, materials which are omnipresent in nature (cellulose, proteins) or in synthetic form (textiles, adhesives). In contrast to metals and other construction materials, polymers are organic materials with properties that depend strongly upon the environment in which they exist. A prerequisite to fully understand the life cycle of a plastic product is to examine the structure and nature of polymers.

The structure of polymers

Polymers are giant molecules, or macromolecules, constructed from smaller repeating chemical units, or monomers. Polyethylene, for instance, is composed of macromolecules created by chemically joining many thousands ethylene monomers, and is represented by the formula $(CH_2)_n$.

$$\cdots -CH_2-CH_2-CH_2-CH_2-CH_2-CH_2-CH_2-CH_2-CH_2- \cdots$$

Polymers are divided in two main classes: *thermoplastics* and *thermosets*, as illustrated in Figure 2.1. In the former, representing 80% of total plastics, macromolecules are not chemically bonded to each other. Thermoplastics become viscous when heated, allowing shaping under pressure, and solidify when cooled. This cycle can be repeated many times. The major thermoplastics in terms of volume are polyethylene (PE), polypropylene (PP), polyvinyl chloride (PVC) and polystyrene (PS). Together these four polymers account for more than 90% of the total consumption of thermoplastics. They are homopolymers, where all monomers are identical, with a variety of configurations or degree of branching. Copolymers, on the other hand, are made from two or more monomer types which can be present in the chain in several fashions: random, alternated, block or grafted. Examples are styrene-butadiene copolymer (SBS) or acrylonitrile-butadiene-styrene terpolymer (ABS).

In contrast to thermoplastics, the polymerisation process for thermosets leads to the creation of a single giant three-dimensionally crosslinked molecule. The crosslinking process is generally irreversible and the resulting material will thus not melt upon subsequent heating. Examples of thermosets are unsaturated polyesters (UP), epoxies and polyurethanes (PUR).

Natural and synthetic rubbers are also three-dimensional crosslinked networks of macromolecules. An example is styrene-butadiene rubber (SBR), obtained by linking SBS molecules with sulphur.

The basic properties of some common polymers are given in Table 2.1.

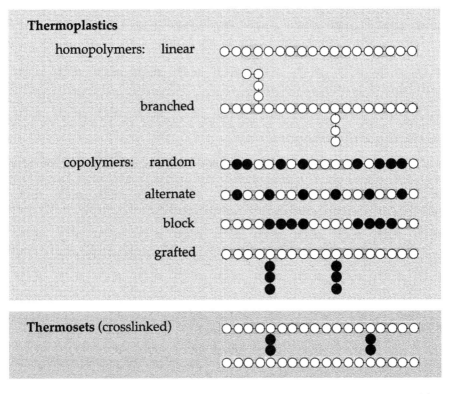

Figure 2.1   Basic classes of polymers (each circle represents a monomer unit).

Table 2.1   Basic properties of common polymers.

Abbreviation	Chemical name	(a) T_g (°C)	(b) T_m (°C)	Maximum crystallinity (%)
Thermoplastics				
HDPE	high density polyethylene	-20	125	90
LDPE	low density polyethylene	-10	115	65
PP	polypropylene	-25	160	75
PS	polystyrene	100	-	0
PVC	polyvinyl chloride	80	-	10
PC	polycarbonate	150	-	0
PA	polyamide 66	60	265	65
	polyamide 6	50	215	0
PMMA	polymethyl metacrylate	105	-	0
PET	polyethylene terephthalate	70	260	40
PTFE	polytetrafluoro ethylene	(+20)	330	75
POM	polyoxy methylene (acetal)	-75	175	85
PEEK	polyetheretherketone	155	400	40
PI	polyimide	440	-	0
Thermosets				
PF	phenol-formaldehyde	*	-	0
UP	unsaturated polyesters *(family of polymers)*			0
PU	polyurethane *(family of polymers)*		-	0
EP	epoxy *(family of polymers)*	120-220	-	0
Rubbers				
NR	cis-polyisoprene	-70	-	0**
SBR	styrene-butadiene copolymer	-55	-	0

*decomposes first, **may crystallise under high tensile stress,
(a) T_g = glass transition temperature, (b) T_m = melting point

Polymers with their randomly coiled and entangled macromolecules can be visualised as a bowl of spaghetti. Beyond this analogy, the structure of polymers is schematically summarised in Figure 2.2. Amorphous thermoplastic polymers have no ordered structure and are therefore inherently transparent. In these materials, including polystyrene, polycarbonate (PC) and acrylates (PMMA), the physical interactions between adjacent chains explain the cohesion of the material. Crosslinked thermosets and rubber polymers are also amorphous. In some cases, linear chains cluster together in local axial alignment to form crystallites within the amorphous matrix, that aggregate into lamellae, and form larger spherulites. These *semi-crystalline* polymers are usually opaque and include polypropylene (PP) and polyethylene terephthalate (PET).

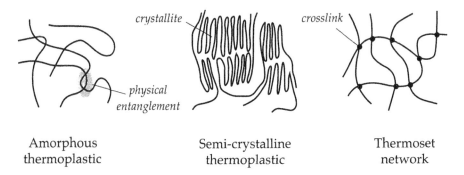

| Amorphous thermoplastic | Semi-crystalline thermoplastic | Thermoset network |

Figure 2.2   Schematics of polymer structures.

Structural transitions in polymers

Upon heating, the properties of any polymer undergo marked changes at certain temperatures, as shown for the elastic modulus in Figure 2.3. At very low temperatures, the polymer is frozen in its glassy state, as no part of the molecules can move. Upon slow heating, the polymer passes through a temperature, often referred to as the *β transition*, at which molecular side groups possess sufficient thermal energy to rotate. This freedom induces a slight drop of modulus. On further heating, the polymer attains its *glass transition temperature*, T_g, at which large macromolecular segments are able to move individually. The modulus drops by typically three orders of magnitude in

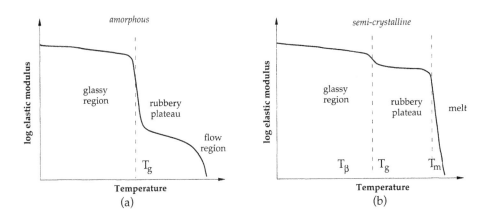

Figure 2.3   Thermal transitions in polymers.

amorphous polymers, whereas the drop in semi-crystalline polymers is significantly smaller and is governed by the degree of crystallinity. Above T_g, the polymer is in its rubbery state. If heated more, an amorphous thermoplastic becomes viscous and starts to flow (Figure 2.3 a), whereas a semi-crystalline polymer (Figure 2.3 b) melts at T_m. At temperatures higher than T_m, the polymer will eventually degrade thermally.

Stiffness and strength of polymers

In structural design, materials are often selected according to their stiffness, strength and durability. Stiffness is commonly expressed through the elastic (or Young's) modulus, E, which is the initial slope of the stress vs. strain curve derived from conventional tensile tests. Strength is the stress at failure. As shown in Figure 2.4, polymers present a large variety of behaviours. The modulus of rubbers is generally close to 1 MPa, and their elongation at failure can be as high as 1000%. Below the glass transition temperature, the moduli of thermoplastics and thermosets are typically in the range of one to several GPa. As already pointed out, the modulus depends on temperature (Figure 2.3). The mechanical behaviour of polymers further depends on strain rate, a consequence of their viscoelastic behaviour, which is described in section 2.4.4.

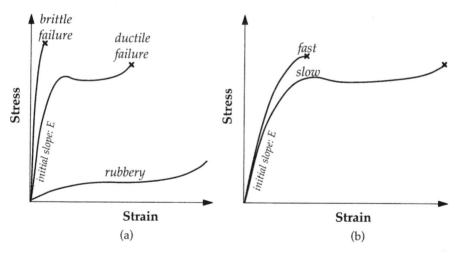

Figure 2.4   Schematic stress-strain behaviour of polymers. A variety of failure behaviours may be observed, which depend on polymer type, structure and/or temperature (a), and strain rate (b).

Depending on their structure and factors such as temperature and strain rate, polymers may exhibit brittle or ductile failure (Figure 2.4). Highly cross-linked polymers well below their glass transition are often brittle, with maximum elongation of the order of 1-4%. Semi-crystalline polymers are also brittle when strained at high rates. On the contrary, at lower strain rates, most polymers are capable of yielding, and their failure is ductile. The Poisson's ratio, v, or lateral contraction ratio, usually ranges from 0.3 for glassy amorphous polymers, 0.4 for semi-crystalline polymers, to 0.5 for rubbers.

Characterisation methods

A wide variety of characterisation methods is available for analysing polymer structure and properties [1-5]. Some of the most frequently-used are listed in Table 2.2, with examples of the information they provide, especially in respect to polymer degradation and ageing.

Table 2.2   Techniques for polymer structure and property characterisation.

Technique	information obtained	information related to degradation and ageing
Infra red spectroscopy	chemical composition	carbonyl content (oxidative stability)
Differential Scanning Calorimetry (DSC)	thermal transitions	oxidation onset time and temperature, melting point, crystallinity
Thermogravimetric Analysis (TGA)	weight loss on heating	thermo-oxidative stability, oxygen absorption rate
Gel Permeation Chromatography (GPC)	molecular weight distribution	molecular weight distribution
Tensile test	tensile strength & modulus, elongation to break	compliance
Creep / relaxation test	time dependent properties	compliance
Fatigue test	fatigue resistance	brittle fracture, crazing, shear yielding
Impact test	impact resistance	reduced impact resistance
Dynamic Thermo-Mechanical Analysis (DTMA)	viscoelastic properties	

2.2 WHAT ARE PLASTICS?

For industrial purposes, where material reliability and durability is required, a large variety of substances called additives have been developed to limit the effect of processing and service conditions on the polymer. The combination of a polymer with appropriate additives creates a plastic.

Plastic = Polymer + Additives

These additives not only prolong the life of the material in its application, but also prevent the material from degrading under the high temperatures and mechanical loads induced by processing. In Tables 2.3 and 2.4 some of the more commonly-used additives are listed.

Table 2.3 Protection against polymer ageing and degradation.

Additive	Function	Substance	Remarks
Heat stabilisers	prevent chain scission, depolymerisation	wide range	some contain heavy metals
Antioxidants	Prevent oxidative degradation	phenols, aromatic amines	can be used to protect material in processing and service
UV protection agents	prevent photo-oxidative degradation	organics, metallic complexes, inorganic pigments	
Flame retardants	inhibit pyrolysis or oxidation reactions during combustion	phosphorous compounds, halogen agents, antimony oxide	especially used in the field of construction
Biocides	hinder or prevent microbiological degradation of the polymer	classified as pesticides by the US Environmental Protection Agency(EPA)	not necessary in many polymers, e.g. PMMA, PE

As will be seen later in this chapter, oxidation and ultraviolet light (UV) attack are the major causes of polymer degradation [6]. All polymers of commercial interest are currently stabilised with antioxidants and UV stabilisers. There are two types of antioxidants, *processing* and *long-term*. The former stabilises the material during processing at high temperatures and is largely consumed during the processing stage, whereas the latter stabilises the material at its service temperature over a much longer time scale. UV stabilisers are necessary for most plastics destined for outdoor use. As oxidation almost always occurs simultaneously with UV degradation the degradation process is accelerated. Several methods are used to hinder UV attack. *External light barriers* can be a paint or a textile, while *internal light barriers* are additives such as carbon black and zinc oxide. The disadvantage with internal light barriers is that they are not transparent and that they are not active on the surface of the part. *UV absorbers* transfer energy of the light into heat instead of bond breaking energy, they allow transparency but are not active on the surface of the part either. *Quenchers* spread the absorbed energy and thus hinder further photochemical reactions, typical examples of quenchers being nickel compounds. The main advantage with quenchers is that their function is independent of the thickness of the material. There is, however, a risk of their discolouring the material. *Radical absorbers* react with free radicals present in the plastic and thus decelerate the chain scission process at the root of the degradation, the best-known being hindered amine light stabilisers (HALS). They are able to protect even very thin layers, and are continuously regenerated so they can be used in smaller amounts than other stabilisers. They can, however, react with other additives.

Table 2.4   Physical property modifiers.

Additive	Function	Substance	Remarks
Plasticizer	decrease viscosity, lower T_g, lower final rigidity	phthalate esters adipate esters	80% used for PVC, the rest for PVA and rubbers
Lubricants and mould release agents	easier processing, improved surface properties	stearates, waxes PTFE, MoS_2, graphite	external and internal lubrication
Macro-molecular modifiers	improve impact resistance, wear, etc.	wide range	grafting, blends, two-phase systems
Reinforcing fillers		e.g. fibres	composites
Reinforcing agents		mineral or organic filler particles, glass microspheres, mica flakes, whiskers	composites
Coupling agents	improved interface by stress transfer, wetting	silanes, organotitanates	composites
Colorants	absorption of incident light	pigments (TiO_2, iron, lead, cadmium compounds), organic dyes, substrates	several banned due to environmental and health risks
Brightening agents	to give fluorescence, phosphorescence, pearlescence effects	organic compounds	
Blowing agents	vaporisation for foaming	chemical: azodicarbonide citric acid based physical: CO_2, pentane	
Antistatics	reduce charge build-up on polymers	carbon black, powdered metals	especially for electronics & food packaging

Lubricants are used to increase throughput in processing. Typical examples of lubricants are metallic stearates and paraffin wax.

Plasticisers, fillers, reinforcements and blowing agents are added to modify mechanical performance. With such additives a wide range of properties can be achieved from the same base polymer. Additives play an important role in the industrial application of polymers since they help ensure performance and stability throughout the service life of the material. They can be used to prolong service life and to aid in recycling with minimal property degradation.

An issue of considerable significance in life cycle engineering is polymer durability and reliability, both for virgin material and (especially) for recycled material. The use of additives is one important aspect of this.

2.3 DURABILITY AND RELIABILITY OF PLASTIC PRODUCTS

While the effects of material and processing variations on polymer properties are quite well-documented [7-10], reliable prediction techniques for the evolution of properties during the service life of a part are still lacking. In a typical application, a polymer component may be simultaneously subjected to mechanical, thermal and chemical stresses such that the characterisation of the durability of such a part is a complex task. Polymers retain memory of preceding life-cycle steps so the durability not only depends on what happens during service, but also on what happened before, during manufacture (Figure 2.5). Hence, the durability of recycled plastics will depend on the full history

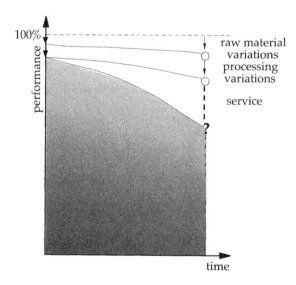

Figure 2.5   Designing with uncertainty–the importance of durability.

of the material. To compensate for uncertainty over long-term polymer properties, over-conservative safety factors are used in designing products, which ultimately translates into wasting resources by over-designing.

Durability is the ability of a material or product to retain performance during a specified time. Reliability is a statistical measurement of the consistency of performance over time. Together the durability and the reliability of a product determine the maximum performance which can be guaranteed during its service life or the equivalent maximal guaranteed service life with properties above a specified limit (Figure 2.6).

The durability and reliability of a polymer-based product are determined by a number of factors inherent to the material itself (for example, crystallinity, average molecular weight), to its processing (for example, shear-induced degradation, process-induced thermal degradation), and to its service environment (such as, temperature, humidity and the presence of vibrations) [11].

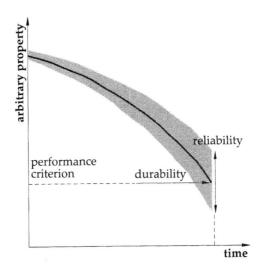

Figure 2.6   The relation between durability and reliability.

To be able to predict durability the following is necessary:

1   Intimate knowledge of the application. It is clear that the first step in determining durability is to specify which performance criteria are vital to the products' function. For example, in an aircraft application such as a structural wing part, performance requirements may be such as retaining dimensional stability and yield stress for 12,000 supersonic hours under cyclic loads and temperatures up to 175° C.

2   An understanding and quantification of the mechanisms of ageing and their relation to the intrinsic properties of the material.

3   Develop time- and geometry-based scaling techniques to relate laboratory tests to the real performance of a structure, and accelerated test methods for simulating long-term ageing.

2.4 DEGRADATION AND AGEING OF POLYMERS

What affects polymers?

Polymer properties change over time. Being organic materials, most synthetic polymers are sensitive to light, oxygen, moisture, heat, and other aggressive environments (Figure 2.7). Long-term exposure to such environments leads to chain scission and other degradation reactions which often results in yellowing, embrittlement, and reduced mechanical properties. Being viscoelastic materials, polymers exhibit time dependent behaviour under mechanical loads. Under a fixed load, polymers creep, which may ultimately lead to failure. They also relax the stresses generated when they are subjected to a permanent deformation. Being induced out of thermodynamic equilibrium by processing operations, crystalline structures may during service evolve towards higher crystallinity, whereas amorphous fractions of polymers slowly densify with time below T_g. Both phenomena will inevitably affect the properties of the polymer.

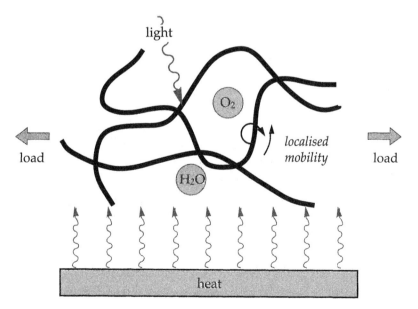

Figure 2.7   Factors contributing to polymer ageing and degradation.

The polymer life cycle

Figure 2.8 shows some of the numerous mechanisms acting independently or jointly to alter the performance of a material during its life-cycle. These mechanisms can be *chemical* in nature or *physical*. The former correspond to degradation effects encountered during processing (such as high-temperature degradation) and during service (low-temperature degradation), while the latter are related to ageing and viscoelastic effects, including for instance the generation of internal stress occurring during the final cooling phase of a processing cycle.

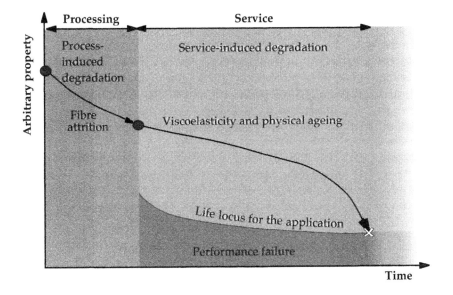

Figure 2.8   Decrease in properties of a composite during processing and service.

2.4.1   PROCESS-INDUCED DEGRADATION

Thermo-oxidative degradation and solvolysis

Oxidation is caused by the presence of atmospheric oxygen, usually in combination with high temperatures. The oxidation process starts with the creation of a free radical in the main polymer chain, as depicted in Figure 2.9. Subsequently, this free radical reacts with oxygen to form oxygenated radicals, which in turn attack the polymer backbone to form a hydroperoxide, while the attacked chain is now a radical that goes through the same process. The oxidation process is accelerated by the presence of metallic ions and is terminated by recombination of various radical groups into inactive products. This phenomena is strongly dependent on the polymer structure: polymers with unsaturated bonds in the main chain, extensive chain branching or low crystallinity are especially sensitive to oxidative degradation.

Thermal degradation is, like pyrolysis, degradation in an inert atmosphere or in the absence of an external reagent. Energy absorption leads to molecular dissociation through three major mechanisms acting alone or jointly:

- random chain scission (in polyolefines),
- depolymerisation (e.g. PMMA, POM, PET),
- elimination of low molecular weight fragments different from the monomer (PVC).

A further process of degradation is solvolysis or depolymerisation. This process is also thermally activated. Chain scission occurs through the action of a solvent on reactive sites on the main macromolecular chain in reactions such as hydrolysis, alcolysis, and acetolysis. Some of these reactions, for example glycolysis, are currently being commercialised as methods of chemical recycling of polymers; polyesters can be chain-cleaved by hydrolysis. PET, for example, can be hydrolysed in acidic, neutral, or basic media [12]. Not only resins are sensitive to solvolysis: stress corrosion of E-glass fibres in the presence of water may also contribute to considerable loss of properties in composites [13].

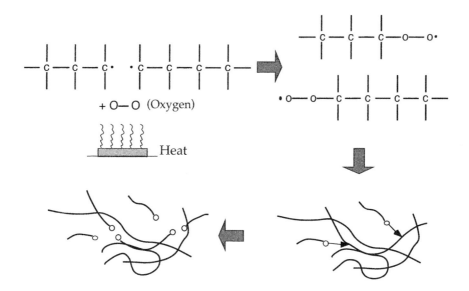

Figure 2.9   The oxidation process.

Flow-induced degradation

Mechanical or flow-induced degradation of polymers may occur under extreme shear or extrusion flow conditions giving rise to unfolding of macromolecules in extensional flow with subsequent chain scission.

In fibre-reinforced materials, other important mechanisms of mechanical degradation are fibre shortening, due to high local shear forces at the interface between the solid bed and the molten polymer, and viscous shear buckling of fibres in the molten polymer (Figure 2.10) [14]. Fibre shortening is a key issue during processing, since the resulting fibre length, fibre length distribution and interfacial properties strongly influence the composite mechanical performance during its service life. This issue is obviously even more important in a situation of closed-loop recycling.

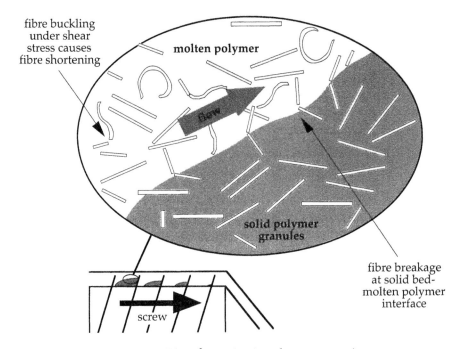

Figure 2.10   Fibre shortening in polymer processing.

2.4.2   SERVICE-INDUCED DEGRADATION

The main features of low-temperature oxidation resemble that of the high-temperature processes described previously. Nevertheless, being thermally activated, low-temperature degradation proceeds at a much lower rate. It is relatively slow in semi-crystalline polymers and in polymers used below their glass transition.

A typical service-induced degradation process is photo-oxidation which usually occurs through a mechanism similar to thermal oxidation. The difference being that initiation is through UV radiation. Photo-oxidation only attacks the superficial layers of a part with serious cosmetic consequences. Outdoor weathering combines thermal and photo oxidation, heat being generated by IR radiation.

Ozone attacks double bonds within polymer chains to form unstable ozonides which further degrade the chains. This type of degradation is predominant for under-the-bonnet applications where ozone formation can be significant.

A further degradation mechanism is solvent absorption which, in addition to provoking solvolysis, also may cause swelling, with significant effects on the mechanical properties of polymers and composites. In composites, Ashbee and Wyatt [15] propose that swelling causes debonding of fibres if the internal pressure thus generated is too high. A further effect of moisture and solvent absorption is plasticising, which lowers T_g, strength, and stiffness of polymers and composites by easing chain movement within the resin. It also increases the elongation to break and retards crack propagation.

It is worthwhile noticing that the above mentioned degradation processes may be affected by process-induced internal stresses. Tensile stresses accelerate degradation, while compressive stresses retard it. In general, the solidification process of polymers generates compressive forces at the surface which slow the ingress of oxidising and solvolytic agents into the polymer.

2.4.3   PHYSICAL AGEING

It has been shown that polymer properties change over time, as a result of the combined action of numerous external factors. The microscopic structure of polymers also changes over time, even in the absence of external factors. Indeed, the arrangement of macromolecules is not fixed, and their mobility is responsible for several property changes.

Upon cooling from above T_g to below it, which happens in most processing operations, several material state variables such as the specific volume of the material depart from equilibrium, as shown in Figure 2.11. At a fixed temperature, T_s, these variables gradually evolve back towards equilibrium, corresponding to a slow *densification* of the polymer [16, 17]. This phenomenon, termed structural recovery, has a profound impact on the performance of all glassy polymers and the amorphous domains of semi-crystalline polymers in that it causes the:

- elastic modulus and yield stress to increase,
- creep compliance and impact strength to decrease,

- coefficient of thermal expansion, dielectric properties and physical and chemical stability to change.

Physical ageing is the term describing these effects [18]. Renewed interest in the physical ageing of polymers has been spurred by newly-found applications for plastics and polymer composites in high-performance areas such as aerospace, electronics, and telecommunication [19, 20]. The lack of understanding of the ageing process and its effects on material properties has long been the cause of economic ramifications. For example, in the photographic film industry, the control of ageing becomes increasingly important as the industry moves towards tighter inventory control and just-on-time delivery procedures. The delivery of unaged or partially-aged film could have grave consequences for the quality and reliability of the product [5].

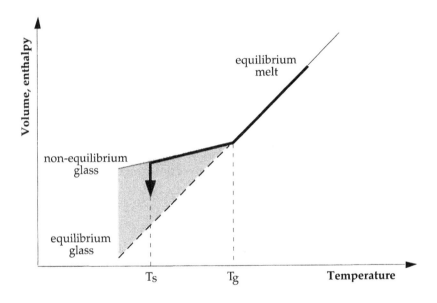

Figure 2.11   Volume versus temperature relationship for the cooling process of a glassy polymer: a slow densification occurs at service temperature, T_s, below T_g.

The evolution of mechanical properties is of evident interest for engineering applications. A decrease in creep compliance or an increase in modulus is generally beneficial for long-term use of materials [21]. In contrast, a decrease in toughness is often detrimental. There are various ways of reducing physical ageing:

- *Proper choice of material*: effects are minimal for polymers with a high T_g and a minor β relaxation in the glassy state. Addition of inorganic filler particles further reduces the effects by reducing the relative density change of the material [22].

- *Cooling under high hydrostatic pressure*: if hydrostatic pressure is applied during the shaping period and then removed after cooling, a densification effect is frozen in that counters to some extent volume recovery during service and in this way lower warpage [22].

2.4.4 VISCOELASTIC EFFECTS

Polymers also demonstrate properties intermediate to that of viscous fluids and elastic solids: they are viscoelastic [23-25]. The most striking phenomenon resulting from the viscoelastic nature of polymers is *creep*: under a fixed load, the material slowly and continuously deforms. A good example of a plastic part exposed to creep is a fan blade in a car; it is subjected to tensile stresses due to high speed rotation and simultaneous heat load. Creep is particularly significant for pure plastics. By adding reinforcing fibres to a plastic matrix or by crosslinking it, tensile creep can be significantly reduced (Figure 2.12 a) [26].

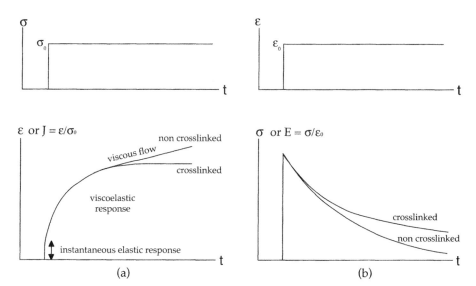

(a)   (b)

Figure 2.12 a   Creep of a viscoelastic polymer subjected to a constant stress, σ_0; the creep response is usually represented by the creep compliance, $J = \varepsilon/\sigma_0$, where ε is the measured strain.

Figure 2.12 b   Stress relaxation of a viscoelastic polymer subjected to a constant strain, ε_0; the relaxation response is usually represented by the relaxation modulus, $E = \sigma/\varepsilon_0$, where σ is the measured stress.

On the other hand, when a part is subjected to a permanent deformation (Figure 2.12 b), a gradual decrease of stress is observed due to *stress relaxation*. Creep and stress relaxation often act together and strongly influence the dimensional stability of plastics-based products: they are of interest to engineers in any application where the material must sustain stress or strain over a long period of time.

Stresses can be either externally applied, which is accounted for at the design step by stress analysis of loading cases, or internally, as a result of processing and part curvature. Reasons for internal stress build-up within a part during processing are:

- flow (frozen-in shear stresses),
- shrinkage (thermoset cure; thermoplastic crystallisation),
- different thermomechanical properties (especially in composites, but also in blends).

2.5 LIFE-TIME PREDICTION

Without proper knowledge of the circumstances in which degradation mechanisms are active and of how they interact, there is no firm base for reliable life prediction models. Products will be over-designed to compensate for the lack of accurate predictions. The models presently available for quantifying the degradation and ageing mechanisms presented in the previous section are reviewed here.

2.5.1 LONG-TERM PREDICTION OF THERMO-OXIDATIVE DEGRADATION

Most degradation processes are temperature-activated, and they are best represented by the classic Arrhenius reaction rate equation. The application of such a model is shown in Figure 2.13. The short-term points are obtained by selecting the life criterion (for example a 50% drop in toughness) and then ageing the material at several elevated temperatures until the desired extent of degradation is achieved. Four such points are recommended. A linear extrapolation on a log(criterion) versus $1/T$ plot allows prediction of the life at

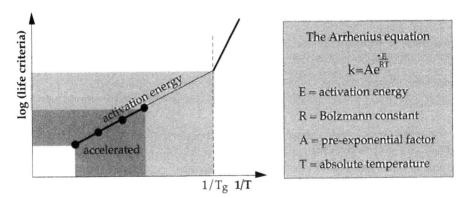

Figure 2.13 Arrhenius-type plot for long term prediction of thermal degradation in polymers. Note that activation energy changes at the glass transition.

lower temperatures. The slope of the line provides the activation energy, E, of the degradation process. Since the activation energy is likely to change with temperature (at T_g for instance), it is generally admitted that this method is limited to a temperature range 25°C beyond the last data point.

It is rare for plastic components to be exposed to a single temperature during their entire service life. An under-the-bonnet component, such as a radiator end-cap, injection moulded from glass-fibre reinforced polyamide 66 could undergo a temperature profile during its service life such as the one displayed in Table 2.5.

Table 2.5   Temperature exposure during useful life of an automotive under-the-bonnet application.

Temperature (T) (°C)	Time at T (hours)	Equivalent time at 200°C ($t_{eq\ 200}$)
140	2000	t_4: 538
110	80,000	t_3: 7523
80	100,000	t_2: 3288
< 25	300,000	t_1: 752

Provided that the energy of activation is known for the base polymer, this data can be plotted in a diagram with the selected life criterion, in this case 50% of the tensile strength of the virgin material, as displayed in Figure 2.14 to obtain a

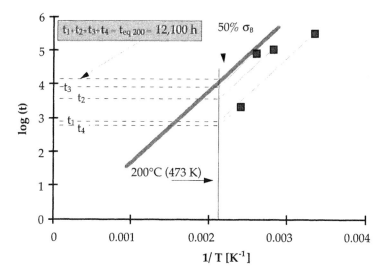

Figure 2.14 Hands-on accelerated testing method for long-term thermo-oxidative degradation.

hands-on prediction method of long-term thermo-oxidative degradation based on accelerated testing. The points corresponding to the time of exposure at a certain temperature are moved parallel to the slope of the life criterion to the temperature at which one desires to perform accelerated testing. By adding the obtained times at this temperature, the required time of accelerated testing at a constant test temperature to simulate the service induced thermo-oxidative degradation is obtained.

2.5.2   SUPERPOSITIONS AND SHIFT FACTORS

Time - temperature superposition

In many applications, plastic parts carry reasonably constant mechanical loads over periods up to few years. The polymer will creep during the lifetime of the part. At moderate load levels, long-term prediction of creep from short-term tests is possible, because the viscoelastic response of polymers (creep, stress relaxation) measured at different temperatures superimpose when shifted along the time axis [24].

The first step is to carry out a series of creep tests at different temperatures, usually above the glass transition temperature of the polymer. A fixed weight is suspended on a sample at a fixed temperature, and the resulting sample elongation Δl is measured as a function of time, usually for less than a couple of hours.

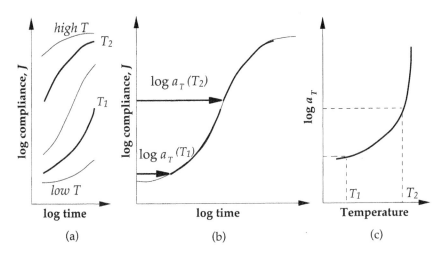

Figure 2.15 Construction of the time-temperature superposition and derivation of temperature dependent shift factor.

The second step is to plot the material compliance $J = (\Delta l\ /\ l_o)\cdot(F\ /\ A_o)$ versus time in logarithmic co-ordinates (Figure 2.15a). l_o and A_o are the initial sample length and cross section and F is the applied load.

The third step is to shift the short-term creep curves measured at different temperatures along the log (time) axis until they superpose one onto another, even partially. This operation defines, for each testing temperature, a shift factor a_T, and produces a master curve (Figure 2.15b). The shift factor is defined as follows:

$$J(T_2,t)= J\Big(T_1,\big(a_T(T_2)-a_T(T_1)\big)\cdot t\Big) \tag{2.1}$$

that is, the compliance measured at a temperature T_2 after a time t is equal to the compliance measured at another temperature, T_1, after the time t multiplied by the shift factor between T_2 and T_1: $(a_T(T_2)-a_T(T_1))$.

The fourth step is to plot the shift factors a_T against temperature (Figure 2.15c). This representation of the *time-temperature superposition* characteristic of viscoelastic materials has been extensively analysed with the well-known Williams-Landel-Ferry (WLF) relationship, at temperatures above T_g:

$$\log a_T = -\frac{C_1(T-T_0)}{C_2+T-T_0} \tag{2.2}$$

Below the glass transition temperature a linear extrapolation of the shift a_T is used. The creep behaviour can then be predicted at any other temperature and time, by combining equations (2.1) and (2.2), provided that no other effects (environmental interactions, chemical or physical aging) occur over the complete time scale.

Time - ageing time superposition

Similarly, the viscoelastic response of polymers at a fixed temperature below T_g measured at different ageing times (Figure 2.16a) superimpose when shifted along the time axis (Figure 2.16b). This *time-ageing time superposition* is expressed through an ageing time shift factor, a_{te}. The linear dependence of this factor on time in a log-log representation defines an ageing rate, μ, which is generally close to unity (Figure 2.16c). The precise value of μ may be derived from isothermal short-term creep tests performed on samples aged at a constant temperature below T_g for various lengths of time.

The prediction of the long-term viscoelastic response of a polymer part subjected to any given temperature history is possible through the integration of

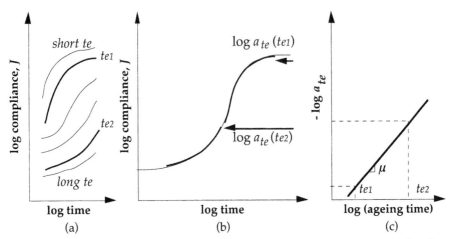

Figure 2.16 Time-ageing time superposition for long-term prediction of volume recovery effects on the behaviour of polymers.

the shift factors a_T and a_{te} into the constitutive law of the material, that is, into the mathematical expression with relates the response of the material to its solicitation (for example creep response to applied load). The presentation of such integration is beyond the scope of the present discussion, and the reader is referred to the works of Brinson and others [20].

It is evident that the various degradation and ageing phenomena discussed in the above sections occur simultaneously. The interactions between these phenomena and the resulting acceleration of property change are subject to widespread investigations, particularly in the field of composites [27].

2.5.3 PREDICTION OF FATIGUE FAILURE

Prediction of fatigue crack propagation is one of the best-known life prediction tools. The fatigue failure process proceeds through initiation and propagation of cracks as depicted in Figure 2.17. In the propagation phase a crack grows by a small amount each cycle. Crack propagation theory relates the number of cycles to failure to stress, specimen geometry, and crack length, and has been the object of extensive research [28-31].

Polymers display several characteristic crack growth rates as shown in Figure 2.18. Often, the nature of the crack surface can give information as to the speed of crack propagation: slow crack growth is characterised by a smooth surface; intermediate crack growth leaves a rougher surface with river lines pointing in the direction of the origin of the crack and fast crack growth which is attained when the fracture toughness of the material, K_{IC}, is surpassed at the crack tip, is generally characterised by a rough surface. Finding the origin of the crack may be very useful in the determination of the cause of failure. The calculation of crack propagation has been covered in several publications [28-32].

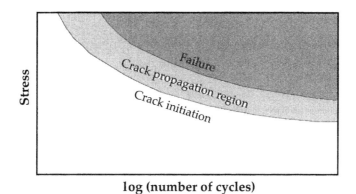

Figure 2.17   Fatigue mechanisms.

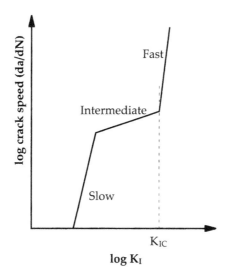

Figure 2.18 Types of crack growth rate for polymers as a function of the stress intensity, K_I, at the crack tip.

Fatigue damage is predominantly caused by crack growth under cyclic loading. Several factors affect the rate at which a crack propagates in a material:

- type of loading (constant strain or constant stress),
- strain energy per cycle,
- value of the median strain (compressive, tensile or neutral),

- temperature (over or below T_g),
- nature of the material,
- nature of the environment,
- loading frequency.

One would expect that fatigue failure would be provoked more rapidly under loads with high strain energy per cycle and at high frequency. Since polymers are poor heat conductors, these two factors may provoke heating by hysteresis which softens the material to facilitate crack growth by crazing. Small faults such as surface cracks may initiate local hot-spots where cracks propagate more easily. Temperature also plays an important role for the propagation of cracks. Lower temperatures makes polymers more brittle and sensitive to crack initiation and crack growth. Finally, the environment in which the material is used can strongly affect the fatigue resistance of the material.

2.5.4   MATERIAL KNOW-HOW AND LIFE CYCLE ENGINEERING

Competitive knowledge of material behaviour and characterisation and predictive skills for long-term performance are vital for the fulfilment of life cycle engineering strategies. Much of the development in life cycle strategies has been concentrated on the field of life cycle assessment and on design support systems for assembly, disassembly and recycling. It will be shown in the following chapters, however, that several important material-related issues still have to be addressed to achieve efficient improvement of life cycle environmental performance.

REFERENCES

1. *Applied Polymer Analysis and Characterisation: Recent Developments in Techniques, Instrumentation, Problem Solving,* J. Mitchell Ed., Carl Hanser Verlag: Munich (1987).

2. *Characterisation of Materials. Parts 1 & 2* in *Material Science and Technology,* E. Lifshin Ed., VCH: Weinheim (1992).

3. *Optical Techniques to Characterize Polymer Systems,* H. Baessler Ed., Elsevier: Amsterdam, The Netherlands (1989).

4. *Polymer Characterisation by Thermal Methods of Analysis,* J. Chiu Ed., Marcel Dekker: New York (1974).

5. C. Arnold-McKenna and G. B. McKenna, *Workshop on Aging, Dimensional Stability, and Durability Issues in High Technology Polymers,* Gaithersburg, MD, May 28-29, 1992, Journal of Research of the National Institute of Standards and Technology, **98**, 4, pp. 523-533 (1993).

6. W.L. Hawkins, *Polymer Degradation and Stabilisation,* Polymers; Properties and Applications, **8**, Springer-Verlag: Berlin (1984).

7. E. A. Muccio, *Plastic Part Technology,* ASM International: Materials Park, Ohio (1991).

8. P. C. Powell, *Engineering with Polymers,* Chapman and Hall: London (1983).

9. J. A. Brydson, *Plastic Materials,* 5th ed., Butterworths: London (1989).

10. S. Levy and J. H. Dubois, *Plastics Products Design Engineering Handbook,* Van Nostrand Rheinold Company: New York (1977).

11. *Durability of Polymer-Based Composites for Structural Applications,* A. H. Cardon and G. Verchery Eds., Elsevier Applied Science: New York (1991).

12. H. R. Allcock and F. W. Lampe, in *Contemporary Polymer Chemistry,* , p. 206 (1981).

13. C.L. Schutte, *Environmental Durability of Glass-Fiber Composites,* Material Science & Engineering: R: Reports, **13**, 7, pp. 265-324 (1994).

14. R. K. Mittal, V. B. Gupta, and P. P. Sharma, *Theoretical and Experimental Study of Fiber Attrition during Extrusion of Glass-Fiber-Reinforced Polypropylene,* Comp. Sc. Technol., **31**, pp. 295-313 (1988).

15. K. H. G. Ashbee and R. C. Wyatt, Proc. Roy. Soc., **312A**, p. 553 (1969).

16. A. J. Kovacs, *Transition Vitreuse dans les Polymères Amorphes. Etude Phénoménologique,* Fortschritte der Hochpolymeren-Forschung, **3**, pp. 394-507 (1963).

17. G. B. McKenna, *Glass Formation and Glassy Behavior* in *Comprehensive Polymer Science, Polymer Properties,* C. Booth and C. Price Eds., Pergamon Press Ltd.: Oxford, UK, pp. 311-362 (1989).

18. J. M. Hutchinson, *Physical Aging of Polymers,* Progress in Polymer Science, **20**, pp. 703-760 (1995).

19. J. Mijovic, L. Nicolais, A. D'Amore, and J. M. Kenny, *Principal Features of Structural Relaxation in Glassy Polymers. A Review,* Polymer Engineering & Science, **34**, 5, pp. 381-389 (1994).

20. L. C. Brinson and T. S. Gates, *Effects of Physical Aging on Long-Term Creep of Polymers and Polymer Matrix Composites*, International Journal of Solids and Structures, **32**, 6-7, pp. 827-846 (1995).

21. L. C. E. Struik, *Physical Aging in Amorphous Polymers and Other Materials*, Elsevier: Amsterdam, The Netherlands (1978).

22. L. C. E. Struik, in *Internal Stresses, Dimensional Instabilities and Molecular Orientation in Plastics*, John Wiley & Sons: Chichester, pp. 17-18 (1990).

23. J. J. Aklonis, W. J. McKnight, and M. Shen, *Introduction to Polymer Viscoelasticity*, Wiley Interscience: New York (1972).

24. J. D. Ferry, *Viscoelastic properties of polymers*, 3rd ed., John Wiley & Sons, Inc.: New York (1980).

25. I. M. Ward, *Mechanical properties of solid polymers*, 2 ed., John Wiley & Sons, Inc.: New York (1983).

26. L. E. Nielsen and R. F. Landel, *Mechanical properties of polymers and composites* , Marcel Dekker: New York, p. 483 (1994).

27. A. H. Cardon, H. Fukuda, K. L. Reifsnider, and G. Verchery Eds., *Recent Developments in Durability Analysis of Composite Systems*, A. A. Balkema: Rotterdam (2000).

28. R. W. Hertzberg and J. A. Manson, *Fatigue of engineering plastics*, Academic Press: New York (1980).

29. W. Brostow and R. D. Corneliussen, *Failure of plastics*, Carl Hanser Verlag: New York (1986).

30. K. L. Reifsnider and Z. Gao, *Micromechanics model for composites under fatigue loading*, International Journal of Fatigue, **13**, 2, pp. 149-156 (1991).

31. S. Suresh, *Fatigue of materials*, Cambridge University Press: Cambridge (1991).

32. P.C. Powell, *Engineering with Polymers* , Chapman & Hall: Bristol, UK, p. 130 (1992).

3

PLASTICS RECOVERY AND RECYCLING

Plastics waste is receiving increased attention in the environmental debate. The technology of recovery and recycling is rapidly progressing, but there is no unique solution for treating plastics waste. The strategy to apply depends on the type of material and product and on consumption patterns. Mechanical recycling, feedstock recycling, energy recovery and environmentally degradable plastics are discussed. The importance of efficient collection and separation systems, of steady material supply and quality and of available markets for revaluated material is shown.

3.1 THE VITAL RECYCLING CHAIN

The recovery and disposal of waste are of growing concern to industry and society alike. They are key elements in what can be termed the "vital recycling chain" (Figure 3.1). Severe restrictions are being imposed on two of the most popular waste management methods, landfill and incineration. Faith is increasingly being placed in the rapidly evolving technologies of recycling, energy recovery and, to some extent, degradable plastics to deal with waste.

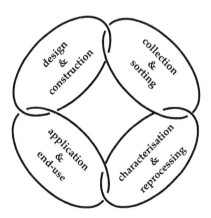

Figure 3.1   Four vital links guarantee successful recycling.

Recovery and recycling are complex issues. Technological evolution has not yet helped to reduce waste, as can be seen from Figure 3.2. In 1994, 17.5 million tonnes of plastics waste was generated in Western Europe. Over one fifth of it was recovered [1] (and of this fifth, 30% was recycled while the remainder was incinerated with energy recovery). Thus approximately 80% of the annual production of a potential source of material and energy is regularly landfilled instead of being recycled or used to replace combustibles. Why all this waste?

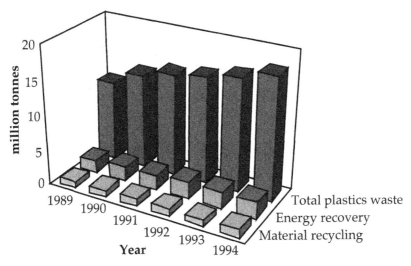

Figure 3.2   Plastics waste recovery in Western Europe 1989-1994 (Source APME, 1994).

Early recycling efforts were plagued by the high capital costs of setting up recycling plants, an irregular supply of material and the low quality of products made from recycled material, which kept recyclers from working at full capacity. Initial applications for recycled material were in low performance products, typically replacing wood. Although these applications divert a fair amount of material from landfill, they are quickly satiated in the long run. Recovery infrastructure and recycling technology need to be improved in order to increase the performance/cost ratio as well as the reliability and durability of recycled plastics. Recycled materials could then be used in a wider range of applications and would represent an asset rather than an expensive burden.

3.2 COLLECTION AND SORTING FOR RECYCLING

In contrast to virgin plastics, used material for recycling emanates from several sources such as municipal solid waste (MSW), automotive dismantlers, moulder's regrind and plant scrap. MSW constitutes the lion's share, but it is also the most problematic material stream to deal with, since it consists of many different materials that may not be easy to identify and separate. Each stream contains material of differing purity and level of degradation.

The efficiency of collection and sorting structures is measured by collection costs, capture rate and purity of generated feedstock. Some current methods of collecting discarded plastic products are displayed in Table 3.1. In kerbside collection households are given directives on at what level they should sort their waste, whereas drop-off and buy-back centres have specifications on what waste types are accepted.

Table 3.1   Collection methods for plastics waste.

Method	Capture rate	Labour/transport costs	Purity of capture
Kerbside collection	70-90% [2] 35-45% [3]	high	proportional to the level of source separation decreases with capture rate
Drop-off centres	10-15% [4]	low	high
buy back centres	15-20% [2]	low	high

Typically, material collected by one of the methods in Table 3.1 is transported by truck to a material recovery facility (MRF), where a second sorting eliminates major contaminants, mixed grades and colours of plastics and unrecyclable material. Thereafter, recyclables are baled and transported to a reprocessor where they are granulated, washed, further separated and repelletised before transport to the end-user.

3.2.1   IDENTIFICATION AND SORTING SYSTEMS

Most collection schemes generate heterogeneous material streams, which must be efficiently sorted to provide generic material streams. Frequently-used sorting methods are:

- *macro sorting* (manual separation of large objects),
- *density methods* (hydrocyclone or float-sink processes),
- *infrared spectroscopy,*
- *selective dissolution and precipitation* (solvent systems).

Macro sorting

Macro sorting can be performed at the source of waste by giving directives to the consumers as in kerbside collection schemes. It is also carried out on an industrial scale at material recovery facilities. Large objects identifiable with a particular application, such as PET beverage bottles, are the simplest to separate in this way. Identification and separation is made easier by the use of logos or bar coding, the best known of which is the Voluntary Plastic Container Coding System of the Society of the Plastics Industry, USA (Figure 3.3). The six specified generic materials cover 95% of domestic plastics waste (Figure 3.4).

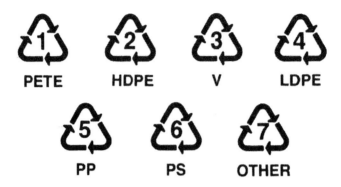

1 PETE = poly-ethylene terephtalate (PET)	5 PP = polypropylene
2 HDPE = high density polyethylene	6 PS = polystyrene
3 V = vinyl/polyvinyl chloride (PVC)	7 OTHER = all other resins
4 LDPE = low density polyethylene	

Figure 3.3   The SPI voluntary plastic container coding system (Reprinted with the permission of The Society of the Plastics Industry, Inc.).

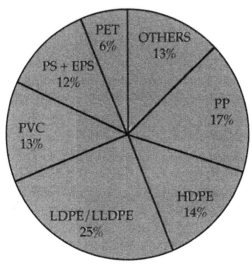

Figure 3.4   Plastics domestic waste by plastic type (wt%) [1].

Such national coding systems may be superseded by the international ISO 14000 standard on material coding (see Chapter 6). There are several drawbacks to these attractively simple systems, of which the most significant is the variety of molecular weights and additives in plastics made from the same base polymer. This can create difficulty in maintaining the required quality of the material stream.

Automated identification and sorting

Automated identification and separation technology is necessary to reduce the cost of recycling and enable greater volumes of waste to be processed.

Density methods; the density of commonly-used plastics varies between 0,9 and 1,7 g/cm^3 [5], but the most common contaminants, such as paper and metal, have densities outside this range. Density-related methods can thus be used for both purifying and sorting plastics. Commonly-used techniques are air classifiers, hydrocyclones, and float/sink baths. Often, air classifiers are used to separate light contaminants such as paper, while hydrocyclones or float/sink baths take care of the heavier fractions.

These methods provide high throughput and sorting efficiency if the feedstock contains only a few materials with reasonable differences in density. They are frequently used for commingled polyolefines. Precautions must be taken to avoid the feedstocks described in Table 3.2.

Table 3.2   Feedstocks not suited for density methods.

Feedstock	Problem
Highly contaminated with dirt and oil	agglomerations of polymers with different densities, dirt and entrapped air make sorting inefficient
Plastics of similar density	immiscible materials may be present in the same fraction
Polymer blends	density does not correspond to a single polymer
Composites / Foams	density does not correspond to the matrix polymer

More efficient density separation is possible by introducing plastic granules into a pressurised vessel containing a supercritical fluid. By altering pressure and composition the density of such fluids can be controlled to within ±0,001 g/cm^3. This allows separation of all polymers with the exception of those with overlapping densities [6]. Productivity remains a problem, however, since this is a batch process.

IR spectroscopy is another frequently-used method of identification of plastics. A specimen is exposed to infrared radiation and the difference between the source and reflected spectra is used to identify the molecular structure of the material. The use of optical filters can improve the speed of the analysis, usually limited by the complexity of the software used to analyse the spectra [7, 8]. The calculations are performed at very high speed. Reported throughput is up to 1 tonne/h when applied to an existing PET processing line [7].

The most frequently-used spectra, in the near infrared range (NIR), cannot be applied to dark-coloured objects such as those common within the automotive sector. This has been resolved by the use of the less readily absorbed mid infrared spectra (MIR).

Selective dissolution is receiving renewed attention as a means to tackle complex waste not suited for density separation or IR-methods such as automotive shredder residue (ASR) and some municipal solid waste (MSW) streams.

In a given solvent, plastics dissolve at different temperatures. A typical process involves the plastics being washed, milled and then dissolved in a heated solvent. By raising the temperature step-wise and using controlled temperature-solvent extraction individual polymers can be separated and then be precipitated by adding a non-solvent. The advantage of selective dissolution is that pure polymer can be recovered without damaging the polymer structure, whereas 2-3% impurities are accumulated during each reprocessing cycle in conventional mechanical recycling. If the composition of the feedstock to be recycled is known, e.g. PET and PVC, a solvent only dissolving one of the two materials can be used to separate the materials from each other [9]. Also, as only cleaning is necessary prior to dissolution, initial separation steps can be avoided. Frequently used solvents are ketones and acetic acids which have low toxicity and degrade quickly.

Selective dissolution can also be used in the recycling of fibre-reinforced composites [10]. Several consecutive washings with a solvent allow the separate recovery of the matrix material and reinforcements [11].

The economic viability of this process, however, will depend on the attainable polymer /solvent ratio. For example, it has been found that under experimental conditions 100g of high density polyethylene (HDPE) may be dissolved in one litre of toluene before the solution becomes too viscous to process [12]. Wietek claims that energy costs for a pilot plant are similar to those of mechanical recycling when running with high polymer/solvent ratios. The financial break-even point for a commercial plant of this type is calculated to be roughly 3,000 tonnes per year [13].

Advocates of selective dissolution claim that the use of large volumes of solvents should not be a problem since these too can be recycled, at levels of up to 95% in experimental trials. Applied on the example above this recycling level would correspond to approximately 1,500 m^3 of solvents to be disposed of per annum. Currently there are no commercial plants in operation due to the high capital costs involved [14].

Molecular markers have been proposed by Eastman Chemical Company and Bayer to readily identify and sort plastics at low cost. The molecular marker is introduced into the base polymer during initial synthesis and allows subsequent identification of post-consumer plastics material by resin type and grade, or by other selected parameters [15].

A similar system working with fluorescent tracer molecules has been developed within a European research project [16]. The material is identified by exposure to an excitation beam, as shown in Figure 3.5, which make the tracers emit their characteristic fluorescent signals. These signals are analysed by the identification system. Subsequently, a sorting machine is activated, which rejects unidentified plastics and sorts other plastics by type.

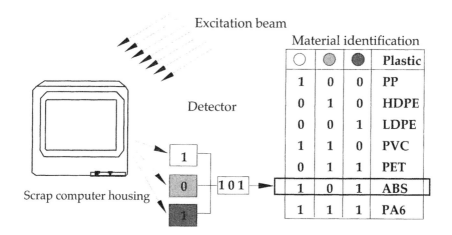

Figure 3.5   Identification and sorting with the use of fluorescent tracers.

There is a practical limit to the number of markers that can be used for each material, as chemical interaction between markers may impede identification of material constituents. Furthermore, surface coating may result in an entire part being identified as consisting of the surface material.

3.2.2   ECONOMICS OF COLLECTION AND SORTING

Reclaiming is labour intensive. The material passes through more hands and is transported longer distances than virgin material. This greatly raises the cost of producing competitive material. The company PET Recycling Switzerland reports return, sorting and compression costs of approximately 1,900 SFr per tonne [17]. Virgin PET flakes sell at approximately 3,000 SFr per tonne, while PET flakes delivered by PET-Recycling Switzerland are sold at 1,400 SFr per tonne [18]. The sale of PET flakes does therefore currently not cover costs. The collection cost for HDPE oil containers from gas stations has been estimated to amount to roughly 185 ECU / tonne, which accounts for 30-40% of the total cost of recycling depending on the intensity with which the recycling facilities are operated [19]. The viable selling price for the generated material is approximately 400 ECU/tonne. which is 150-300 ECU below the selling price required for 10% profits.

As the cost of collection, sorting and repelletizing is largely independent of the type of plastic, it is clear that the initial price of a virgin plastic strongly influences the economics of recycling. At equal collection and sorting costs recycling of HDPE beverage bottles generates resins close to virgin prices, while there is still a possibility for a margin with PET beverage bottle recycling.

Approximately 40% of the total plastics consumption of Western Europe is used in packaging. Of this, approximately two-thirds consist of thin films, extrusion coatings, and foams [20]. These products are not well-suited for recycling, as they are difficult to separate from the municipal waste stream. Their recovery and recycling could consume far more energy than would be saved by their recycling [21]. In commingled recycling, additional transportation steps have multiplied recycling costs by up to ten compared to homogeneous recycling of well defined sources such as PET bottles [22].

A number of criteria need to be satisfied to attain economically viable recycling of post-consumer plastics:

- *Energy-efficient collection:* transport contributes heavily to the energy consumption and cost of recycling operations, especially for commingled waste recycling.
- *Fast collection:* the sooner material returns for recycling from use, the lower the risk of environmental contamination.
- *High capture rate and throughput:* to obtain a sufficiently large and stable material supply to sustain markets.
- *Efficient cleaning and materials separation:* allows producing material of higher quality and performance corresponding to relevant standards and criteria.
- *Low-cost operations:* to give the recycled material a competitive edge.

Companies can resolve their sourcing problems by developing their own infrastructure for collection. These collection schemes typically involve customers who supply clean and relatively homogeneous material which can be made into useful products [23, 24]. One example is the DuPont Partnership for Carpet Reclamation, which recovers nylon carpets from carpet companies for mechanical and chemical recycling. This scheme generates material approved for under-the-bonnet automotive applications.

3.2.3 CONCLUSIONS

Each of the described collection and sorting methods have inherent strengths and weaknesses. They will most likely have to be used in combination to achieve good sorting. It remains to be seen whether the economy and energy efficiency of such systems will be competitive compared to incineration or landfill. As will be seen, the efficiency of this first step of plastics recovery determines not only the way the material will be recycled but also the applications for the generated feedstock.

> "The cleaner the material, the easier it is to recycle and the more it is worth."

3.3 WASTE MANAGEMENT ROUTES

There are currently three viable alternatives for the recovery of plastic waste: mechanical recycling, feedstock recycling, and energy recovery (Figure 3.6). Mechanical recycling consists of the melt reprocessing of solid plastics with conventional processing equipment. Feedstock recycling is achieved by dissociating plastics chemically or thermally into monomers or macromolecules which can then be repolymerised into neat polymer or turned into fuel. Energy recovery by incineration occurs by incineration of solids or by an intermediate conversion to liquid fuel.

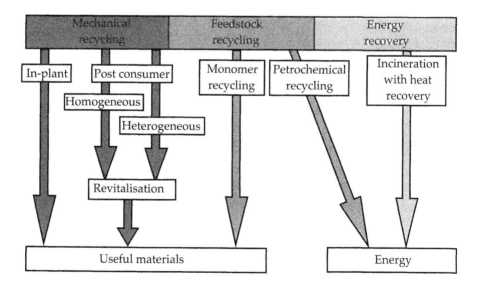

Figure 3.6   Recovery routes for plastics.

The purity and quality of the feedstock determines the appropriate route. With current technology and recovery infrastructure, mechanical recycling is suitable for roughly 15-20% of the annual waste production within the automotive and electrical/electronic sectors, for reasons of quality, markets and economics [25]. To meet EEC recycling targets, European industry organisations and governments are promoting Integrated Waste Management (IWM) strategies which ensure that environmental consequences are considered when evaluating recovery routes.

3.3.1 MECHANICAL RECYCLING

Mechanical recycling provides the highest potential for economy of energy since energy consumption is low, and the energy initially used to synthesise the polymer is conserved. It is the reprocessing of previously-processed materials and involves the use of additives to compensate for property loss during service and reprocessing. Two forms of mechanical recycling can be distinguished: in-plant recycling of previously-unused material and post-consumer recycling.

In-plant recycling

Runners, sprues and off-specification products are common in production. The materials involved are easily identifiable and are of high quality. Instead of being rejected as waste, they can be blended in with virgin material to give acceptable products, provided that certain regrind contents are not exceeded, and that the flow of reground material is sufficiently well managed to avoid mixing with other materials. In-plant recirculation is already a commonly-employed cost-minimisation method. The packaging industry is heavily dependent on satisfactory in-plant recycling levels to maintain low prices in products made from commodity plastics.

If material is continuously recirculated within the same product line, small fractions in the material will undergo several reprocessing steps. If recyclate content is small, typically a couple of per cent, this proportion will rapidly decrease. For smaller products, where runners may have the same volume as the product itself, however, significant multiple-processing degradation might occur and may result in premature degradation of a product.

Knowing the percentage of regrind added to each processing cycle, the proportion of material having gone through any given number of cycles can be determined as shown in Figure 3.7 for regrind levels between 10 and 90%.

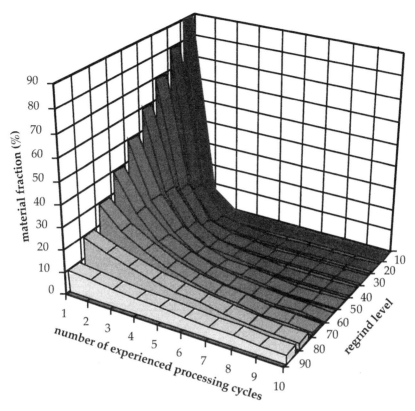

Figure 3.7   Proportions of material moulded different number of times using different levels of regrind in a continuous recycling operation [26].

Within the field of engineering thermoplastics, in-plant recycling has only recently been adopted due to stringent performance specifications. There are no general guidelines on maximum regrind content. Feasible levels depend on the application requirements and on material characteristics.

Post-consumer recycling

Post-consumer recycling still remains the greatest challenge and opportunity for the plastics industry. In comparison with in-plant recyclate, post-consumer waste (PCW) undergoes degradation of the polymer during service and is likely to be contaminated. Homogeneous PCW can be reprocessed in much the same way as in-plant recyclate, whereas commingled PCW requires special processing techniques since municipal solid waste, the dominant source of commingled PCW, commonly contains several mutually incompatible polymers.

A generic recycling flow scheme for homogeneous PCW is depicted in Figure 3.8. In contrast to processing of virgin material and in-plant scrap, post-consumer material requires intermediate stages of decontamination from metal residues and other contaminants. Waste products are reduced in size by a primary grinding. Following contaminant separation a secondary grinding step further reduces the size of the particles before they are cleaned and dried.

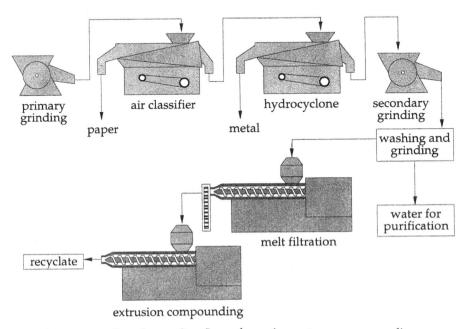

Figure 3.8   Generic recycling flow scheme for post-consumer recycling.

Extrusion with melt filtration can then be used as the ultimate way of purifying the material. Since each extrusion step exposes the material to thermomechanical cycles likely to further degrade the material, the number of extrusion steps should be kept to a minimum. For the final extrusion compounding, additives can be used to maintain or improve selected material properties, depending upon the composition and degree of degradation of the material.

A significant fraction of plastics and products cannot be identified and recovered in a cost-effective way as individual polymers. This includes:

- multilayer packaging,
- composite items made from plastics and one or several other materials,
- highly contaminated items,

- highly pigmented or filled generics,
- different molecular weight grades of the same polymer,
- non-recognisable plastic items,
- minor amounts of specific polymers,
- waste from recovery operations.

Recycling of commingled PCW presents a number of additional problems to that of melt reprocessing of generic plastics from post-consumer or in-plant regrind sources, such as contamination, materials with different melting index and otherwise immiscible materials. Such combinations often dictate very delicate, or almost impossible, processing conditions. As an example: the presence of material with high melt index in commingled feedstock would require mixing at high shear rates and high temperatures. Any heat-sensitive polymers will be degraded under such conditions. The lower temperatures required to avoid such degradation would be insufficient to melt the high melt index polymers [27].

In the past, commingled PCW has commonly been reprocessed using standard techniques such as extrusion, injection moulding [28] or compression moulding [29], but only to produce parts of thick cross-section. In this way small imperfections have a limited effect on mechanical properties and colour variations and surface roughness can be tolerated. Typical applications have been lumber replacement and road markers. Several processes have been developed for the processing of post-consumer commingled feedstocks with the objective of generating high-quality material without the need for tedious and expensive identification and separation. The NewPlast process [30] is based on intensive mixing (homomicronisation) in a cylindrical drum that rotates at high speed. The process is claimed to push the constituent plastics beyond their normal limits of incompatibility. A thermoplastic material has been developed based on HDPE, LDPE, PP, PET, and PVC, that shows higher stiffness than LDPE within the temperature range of 0°C - 85°C combined with significant elongation at break. The material is suitable for processing by conventional machinery. Targeted applications are within the agricultural, construction, houseware, and automotive sectors.

Another process is the Solid-State Shear Extrusion pulverization process [31] developed at the Polymer Reclamation Center at BIRL, Northwestern University, USA, in which unsorted commingled post-consumer waste is subjected to high shear and pressure. By rapidly removing heat from the process, the polymer mix is transformed to a uniform powder of controlled particle size. The process treats polyolefines such as PET, PS, and PVC. The resulting powders have homogeneous and often light colours and are suitable for conversion into products by any thermoplastic processing technique. No segregation has been detected upon blending and complex part have been moulded successfully.

3.3.2 REVITALISATION

Processing operations and the conditions encountered during service life tend to diminish the properties of the materials contained in a product. The chemical nature of the base polymer can undergo significant modification. Additives and stabilisers are added when the plastic is first manufactured, but these are consumed during processing and use, so that when the product is discarded the plastic will not necessarily be reusable. Some common problems encountered during reprocessing and the responsible mechanisms are shown in Table 3.3.

Table 3.3   Problems of reprocessing.

Problem	Source	Occurrence
Low melt viscosity: difficult to extrude Brittleness: demoulding becomes difficult	Polymer chain shortening	Processing
Thermal degradation during processing	No process stabilisers	
Degradation during processing if temperature too high Unmelted regions during processing if temperature too low	Immiscibility —> different melt rheology and melting point	
Difficult to produce thin-walled parts	Impurities	
Brittleness High melt shrinkage Low dimensional stability	Polymer chain shortening	Service
Rapid yellowing Low chemical resistance Surface embrittlement Poor electrical properties	Insufficient antioxidants	
Accelerated deterioration of physical properties Yellowing	Insufficient heat stabilisers	
Poor surface appearance Colour variations	Immiscibility	
Poor surface appearance Voids Embrittlement	Impurities	

Appropriate revitalisation requires knowledge of:

- the effects of the previous life-cycle(s) on material properties,
- material characterisation methods,
- performance requirements for and expected life of the secondary application,

• processing and compounding techniques.

Recent years have seen considerable advances in these fields. The current most-frequently used revitalisation technologies are compatibilisation/ blending and restabilisation.

Early recyclers were often entrepreneurs applying existing technology in a new field. Thus conventional additives are still extensively used in revitalisation. However, recyclers are increasingly cooperating closely with additive and resin producers to develop new products specifically intended for use in plastics revitalisation. As material property requirements for the end-use of recycled materials are still only vaguely defined, few products have yet entered the market. Further work will be necessary before resin and additives manufacturers fully focus their research on means to improve recyclate properties [32].

Blending and compatibilisation

Blending and compatibilisation is a traditional method for modifying the properties of virgin plastics to obtain specific properties. It is financially attractive when compared to the development of new polymers; less R&D and capital investment is required and the time to market is shorter. It was estimated in 1987 that 60-70% of polyolefines and 23% of other polymers were sold as blends [33]. Experience gained when blending virgin polymers can be applied to recycling. Recycled blends are offered by several of the large material producers for applications within the automotive sector, packaging and leisure equipment among others [34, 35].

There are two main categories of polymer blends, miscible and immiscible (Figure 3.9). Miscible blends are homogeneous down to the molecular level [33]. Properties scale with the ratios of the constituent polymers [36].

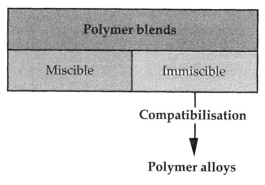

Figure 3.9   Polymer blends and alloys.

If a blend only fulfils the criterion of negative free energy of mixing it is said to be partially miscible. Immiscible blends are frequently heterogeneous and mechanical performance is often inferior to that of the individual constituents in terms of impact strength, elongation at break and resistance to stress cracking. Immiscible blends can often be compatibilised by several means of which the addition of graft and block copolymers in small quantities is the most relevant. These compatibilisers strengthen the physical and chemical bonding between the different phases or decrease the surface energy of the interface. A popular choice of compatibilisers are block copolymers containing chemical groups of the same nature as the polymers in the blend. These interact chemically of physically with the polymers to create a bond between them. Typical examples are such as styrene-butylene molecules, poly(vinyl methyl ether), poly(styrene ethylene), and EPDM elastomer [33, 37-39]. In several commercial blends, modifiers consisting of a copolymer containing a rubbery component are used. These perform two functions: compatibilising and toughening the blend. Examples of such modifiers are thermoplastic polyurethane elastomers, polyacrylate rubbers and polyester ether elastomers [33].

Figure 3.10(a) shows a typical structure of immiscible constituents on the fracture surface of a PP/PET blend [40]. Droplets of PET are embedded in a PP matrix. Since the surface energy of the interface is unfavourably high, the PET assumes the highest possible volume/surface ratio. The lack of PP residues on the surface of the PET particles indicates low adhesion between the two phases. As a consequence, although the individual constituents are ductile, their weak interface promotes debonding and low ductility.

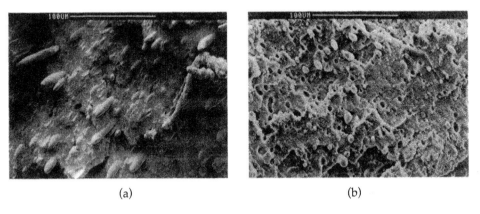

(a)   (b)

Figure 3.10   (a) Immiscible zones of PET in a matrix of polypropylene; (b) PP-MAH compatibilised PET / PP blend.

The addition of 5% by weight of maleic anhydride (MAH)-grafted PP to the material in Figure 3.10(a) improves the interaction between the two phases. The PP component of this molecule strongly interacts with the PP matrix, while the

MAH component interacts with the PET, resulting in lower interfacial energy. This allows a finer dispersion of the PET domains (Figure 3.10(b)), and improvement of the adhesion between the two phases [41].

The optimal percentage of PP-MAH resulting in dramatically improved ductility was found to be 10 wt% [42]. The general effects of a compatibiliser are the following [43, 44]:

• increased impact strength, elongation at break and stress cracking resistance;
• decreased stiffness, hardness, and heat resistance.

Immiscibility in not a phenomenon that is exclusive for heterogeneous mixtures of different plastics; even plastics of the same base polymer may be immiscible with each other, depending on the molecular structure and size as well as on the presence of additives.

High stress shearing techniques, such as those discussed in the section on post-consumer recycling, are in most cases still at an experimental stage. They are known to generate three-dimensional intertwined networks of polymer strands that demonstrate significant thermal stability [33].

Restabilisation

The aim of restabilising is to prepare a material for a new service life. It is first necessary to establish the content and nature of any residual short- and long-term stabilisers present, as well as the effects of processing and use on these stabilisers. The type and quantity of stabilisers that must then be added will depend upon the application requirements. To calculate the effect of repeated processing and use of a material, repeated-processing with intermediate ageing can be done [45].

Stabilisers increase the cost of the material, however. The use of regenerative or slowly-consumed stabilisers can help to keep revitalisation costs down.

Other additives

A frequent problem in repeated processing is chain scission. It result in a reduction of melt viscosity thus making extrusion difficult, an increase in the brittleness of the material making demoulding after injection moulding difficult, and an increase in melt shrinkage (Table 3.3). Furthermore, the final properties of the material in terms of dimensional stability, impact resistance, chemical resistance, and electrical properties are adversely affected. Several additives have been developed as remedies. Some adjust the melt viscosity to desired levels by vis-breaking the polymer chains. Vis-breaking is controlled degradation of a polymer to produce a narrower molecular weight distribution which improves melt flow properties [46]. Others increase the tensile strength and elongation of recycled grades by crosslinking without creating gels [47]. A further effect of such modifiers is to facilitate the grafting of monomers onto the polymer backbone during reactive processing.

Another method of improving properties is to add inorganic fillers or reinforcements. If the designated secondary application for a recycled material requires higher rigidity, addition of reinforcements will improve the properties. A new material is created with different mechanical properties and in which smaller impurities do not have such a strong influence as in a pure plastic.

Environmental aspects of additives

Many commonly-used additives have been linked to environmental and health and safety problems. Table 3.4 lists known problems with several classes of additives.

Several of the substitutes have performance or cost disadvantages. Some of the early problems have been overcome, while in cases such as that of blowing agents the market has been forced to change following prohibition of the production and use of CFCs by international agreement in 1996 [48].

Table 3.4   Environmental problems with traditional additives.

Additive	Standard type	Problem	Replacement
Physical blowing agents	CFC, HCFC	Ozone depletion	cyclopentane, CO_2, chemical blowing agents
Heat stabilisers	cadmium, lead, barium compounds	Toxic leaching in landfills & incineration	Calcium-Zinc, Barium-Zinc compounds
Colorants	cadmium, selenium, lead, chromium compounds	Toxic leaching in landfills & incineration	organic compounds
Flame retardants	brominated compounds, antimony synergists	Toxicity, carcinogenic (dioxins and furanes)	phosphorous/brominated compounds, aluminatrihydrate (ATH)

Colorant replacements have performance and cost disadvantages, especially in durable engineering applications [49]. The same goes for several flame retardant replacements such as ATH and magnesium hydroxide which still suffer performance disadvantages sometimes requiring loadings up to 65% to be efficient [50].

The German Ministry for the Environment is targeting a ban on halogenated flame retardants by the year 2000 [50]. Similar movements concerning environmentally hazardous substances can be seen in governments world-wide. The potential ban on such additives has already changed the preferences of some end-users. Although some of the replacement products still suffer from competitive disadvantages, there is a large thrust behind their development and more performant and environmentally benign alternatives can be expected.

Conclusions for mechanical recycling

It is often technically possible to restore physical properties to a recycled heterogeneous material. It is, however, not always economically justified, since the added cost may take away the profit margin [27]. Since the properties of mechanically-recycled mixed plastics will necessarily be inferior to those of virgin material, it is better to perceive them as *new materials with their intrinsic characteristics*. Progress in separation, cleaning and up-grading technology is likely to improve the consistency and quality of heterogeneous recycled material and encourage their use in more demanding applications.

Confidence for recycled material has been limited among design engineers due to a lack of accurate life performance prediction tools. Often only small quantities are accepted in demanding applications unless there exists legislation enforcing a minimum recycle content. Many producers of in-plant and post-consumer recycled material claim that it is often possible to use larger quantities; the problem rather being to convince material customers that this does not represent an undue risk.

3.3.3 FEEDSTOCK RECYCLING

Feedstock recycling is a new route for converting plastics waste by returning it back to its original constituents, that is, monomers or petrochemical feedstock. Two types of feedstock recycling can be distinguished (Figure 3.11): chemical processes and thermal processes.

Depolymerisation by chemical means is called solvolysis and produces monomers and oligomers. The chemical reactions used to create condensation polymers such as PET, PA and PC can be reversed to convert them into their

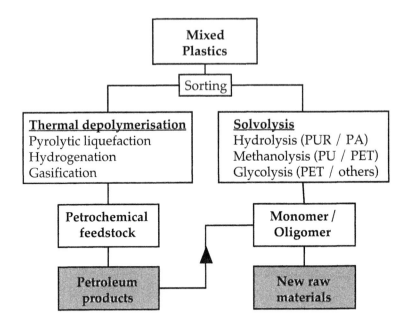

Figure 3.11   Feedstock recycling pathways.

original constituents. The thermal decomposition of polymers is called thermolysis and produces liquid and gas fractions of petrochemical products. Polymers such as PE, PP, PVC, and PS created by irreversible reactions can only be depolymerised by thermolysis. Certain condensation polymers, such as PMMA, can also be recycled by thermolysis [51].

Solvolytic processes

Solvolytic processes are essentially depolymerisation processes involving reactions with water, alcohol or other solvents. These processes are known under various names such as hydrolysis, methanolysis, glycolysis, and ammonolysis. Condensation polymer scrap is heated in presence of a liquid and the polymerisation process is reversed. Such processes are generally intolerant of impurities: addition polymers cannot be recycled this way and must thus be carefully extracted beforehand. These processes are best suited to easily defined large volume sources of relatively pure material [52]. PET has been the target of many depolymerisation research efforts. Already in 1991 both Coca Cola and Pepsi Inc. reported the production of beverage bottles containing up to 25% recyclate obtained by methanolysis and glycolysis, respectively [53]. A plant for chemical recycling of mixed polyamide-6 and polyamide 66 carpet waste by ammonolysis is scheduled for construction in 1998-1999 by DuPont de Nemours. PA 66 has already been successfully produced from hexamethylene diamine (HMD) derived from old carpets. Spun into bulk filament carpet yarns, this material has been used to produce new carpet [24].

Thermolysis

The three major thermal depolymerisation processes are pyrolytic liquefaction, gasification and hydrogenation. If heat is applied to waste plastics in the absence of air the process is called pyrolysis; if done with a controlled oxygen flow it is called gasification. Hydrogenation is a modification of the refining process for petroleum.

Pyrolytic liquefaction produces liquid precursor products or synthetic crude oil that are suitable as refinery feedstock for new monomers. Further products from this process are non-condensable gas, which is used to provide process heat, and solid residues in the form of char.

Mixed polymer streams can be handled in addition to a certain level of non-plastic contaminants, which improves the economy of the process. Even paper and food can be pyrolysed simultaneously to give other chemical products. Consequently the method is ideal for commingled plastics and composites [25]. An additional advantage is that when contamination can be reduced to a minimum, both the matrix and the reinforcing material can be recovered from polymer-based composites.

Gasification involves the partial oxidation of various hydrocarbons into a synthesis gas consisting of CO_2 and H_2 that can then be used for the production of ammonia, methanol, and other types of alcohol [54]. The process takes place at higher temperatures than pyrolysis (1600°C). If the gases are separated, the CO_2 and H_2 can be valuable as chemical intermediates, with two to three times the fuel value of the mixture. In co-operation with Voest Alpine, Dow Europe is developing a high temperature gasification process which converts mixed shredded plastics into a clean fuel, suitable for use in combined heat and power generation [55].

Liquid fuels can be produced by carbon hydrogenation. Plastics waste is ground into small pieces and introduced into a reactor where it is depolymerised by heat at high pressure in an excess of hydrogen [52]. The result is a high-quality petrochemical feedstock that can be cracked into saturated hydrocarbons and syncrude. This is in turn is used in the synthesis of new polymer. The advantage with this process is the ease of separation of side products such as heavy metals, sulphur and chloride. Furthermore existing oil refinery units, such as thermal and catalytic crackers, can be used to convert commingled plastics into crude oil fractions [56]. The advantages of this are:

- capital costs are lowered since new units do not need to be constructed;
- the high hydrogen content and similar chemical structure of most plastics make them easy to integrate with the existing feedstock in refineries;
- the petrochemical feedstock obtained in the refineries can be used in an already established market.

As most refineries are designed for liquid products, then plastics waste has to be converted to a liquid before treatment in a refinery.

Conclusions to feedstock recycling

Feedstock recycling is attractive for obtaining virgin-grade material from plastics waste. In returning to the monomer state, however, the performance value of the polymerised material is lost. Depolymerisation is energy intensive and this accounts for the high cost of the final material.

Due to its sensitivity to contamination from other polymers and from impurities, solvolysis will most likely find application within well-defined product areas with efficient collection and sorting infrastructures. Thermolysis, on the other hand, shows high potential for treating mixed plastics waste such as municipal solid waste and automotive shredder residue which otherwise would go to landfill. Not all thermolytic processes are suited for the generation of new polymers. and a certain amount of the oils, gases and solid residues from these processes is used to replace fuel.

Most feedstock processes are still at the development or pilot plant stage, due to low feedstock supply and high capital and operation costs. Infrastructure and organisational improvements to guarantee a stable supply and markets for the generated material are required for such costs to become acceptable.

3.3.4   ENERGY RECOVERY

Traditionally, incineration has been used to reduce waste volume. It has also been used to produce inert residuals from hazardous waste. Emissions in the form of combustion gases and solid residuals from conventional incinerators have lead to much resistance to the widespread use of incineration. Nowadays, incineration technology is available which avoids the production of such emissions [25]. It would seem that the use of clean combustion technology to recover energy from waste otherwise destined to landfill has been accepted by European governments in and attempt to reduce landfill [57]. Two of the applications for plastics waste under investigation are the co-combustion of plastics waste with municipal solid waste as the main fuel and the use of plastics as a substitute for coke in the operation of cement kilns. If economical or technical considerations prevent mechanical or feedstock recycling, energy recovery should be considered.

Co-combustion of plastics waste with municipal solid waste

The energy content of plastics is comparable to that of ordinary combustibles such as diesel oil (Figure 3.12). It has been shown that the addition of between 7.5 and 15 wt% of plastics waste to municipal solid waste improves combustion in the gas phase as well as of the solid residues due to a more stable and intensive combustion zone [58]. It also reduces waste volume by two thirds [59]. The mineralised combustion residues need to be disposed of as hazardous waste.

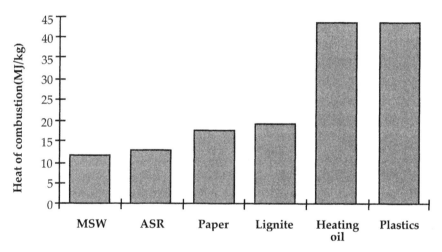

Figure 3.12 Comparative chart of the heat of combustion of various substances on incineration.

There are, however, limits to the allowable energy of combustion in municipal waste incinerators (Figure 3.13). To accommodate increasing energy content in waste, the throughput of incinerators must be lowered, which defeats the aim of increasing the use of incineration in Switzerland: at the current rate of increase of plastics in MSW, incinerator capacity is projected to decrease to 60-65% of today's value by the year 2000 [60]. One possible solution is to convert by pyrolysis the polymer content of MSW into liquid feedstock, which could then be burnt as fuel (or refined chemically into virgin polymer if the quality of the feedstock permits).

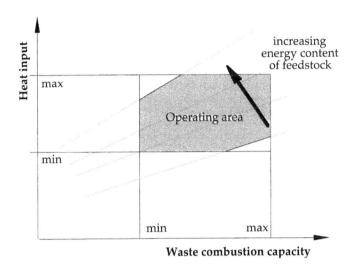

Figure 3.13   Schematic of process operating diagram of an incineration plant (after Nüesch [61]).

Use of plastics waste as a fuel substitute in cement kilns

Cement production involves high temperature combustion in alkaline conditions. It is possible to replace fuel and raw material with residues from other industrial sectors. Scrap tyres are already used to supply up to 20% of the total fuel requirements [62]. The Swiss cement industry is planning to replace 75% of its thermal energy requirements with alternative fuels such as plastics waste. Emission regulations and product requirements limit the type of plastics waste that can be accepted to products such as agricultural mulch film, plastics packaging and process waste from the plastics recycling sector which are low on heavy metals and halogenated compounds [63].

Conclusions

Energy recovery is indispensable if the problem of landfill is going to be solved for materials which cannot be recovered by recycling. Ecologically-leading countries such as Switzerland, where energy is recovered from 72% of the municipal solid waste, and Denmark, where 60% is recovered [64], are leading the way towards the acceptance of energy recovery as an alternative to landfill. In France, a study has concluded that energy recovery from the plastics in shredder residues alone would save annually 120 000 tonnes of crude oil [65] which corresponds to approximately 50% of the daily crude oil consumption in France 1992.

Policies on incineration vary from country to country, for several reasons. First, the capacity to incinerate waste in an environmentally-acceptable manner varies, as does the capacity to deal with plastics waste by mechanical or chemical recycling. National heat demand and supply, fuel prices and taxation also plays a decisive role in the choice between landfill and incineration [66]. Energy recovery should not be seen as a solution to avoid recycling plastics, but rather as a last alternative to landfill should other routes to revaluation prove not to be viable.

3.3.5   ENVIRONMENTALLY-DEGRADABLE POLYMERS

In the context of plastics waste management, environmentally degradable polymers should also be mentioned [67]. As a complete solution to the waste management problem they were perhaps overpromoted at first. Unsatisfied expectations and unproved claims on degradability have created serious scepticism towards these materials [68].

Applications

Despite early drawbacks, environmentally degradable plastics have found a number of applications ranging from agriculture to surgery, cutlery, plates, and cups for single use [69]. Particulate starch-based products, biopolyesters and gelatinised starch products are interesting examples [70, 71]. These types of material have properties resembling those of polyethylene but are several times more expensive. Other materials under development are lactic acid based products [72], aliphatic polyesters [73], and poly-R-3-hydroxy alkanoates. Some materials already approach the price and performance of volume thermoplastics such as PET and PP [74].

Standardisation

In parallel to industrial developments, standardised test methods and definitions are required to characterise degradation and mechanical performance. By definition, biodegradation of a plastic is a *process leading to a change in its chemical structure caused by biological activity leading to naturally occurring metabolic end products* [75]. The ASTM [76] has defined:

- *Degradable polymer*: a polymer designed to undergo a significant change in its chemical structure under specific environmental conditions, resulting in a loss of properties that may vary as measured by standard test methods appropriate to the polymer and the application in a period of time that determines its classification.

- *Biodegradable polymer*: this is a degradable polymer in which degradation results from the action of naturally-occurring micro-organisms such as bacteria, fungi, and algae.

- *Hydrolytically degradable polymer:* a degradable polymer in which the degradation results from hydrolysis.

- *Oxidative polymer:* a degradable polymer in which degradation results from oxidation.

- *Photodegradable polymer:* a polymer in which degradation results from the action of natural daylight.

Confusion of these terms has contributed to exaggerated optimism as to the value of these materials in waste management.

Fate and effects of degradation products

Biodegradable polymers are perceived by the public eye as being environmentally safe, as they would ultimately *disappear* after use. This of course is never the case. Deterioration and degradation of organic polymer structures are the basic mechanisms ensuring recycling of the elements of the biosphere. All organic polymers are, in principle, degradable, but even nature combines inert and degradable components in the same material, as, for example, in wood. The idea that synthetic polymers should biodegrade totally in periods of a few months or less does not reflect the functioning of the natural environment [77].

Polymers degrade through chemical reactions that change the nature of the macromolecules, through reduction into plastic dust, and through both mechanisms acting simultaneously. It is important to determine the effects such degradation products have when they enter the soil or water. Will they be assimilated and reinserted into biological cycles, to be converted to water and carbon dioxide; will they be dispersed and diluted without further interaction, or will they have a negative effect on the ecosystems in which they are present?

Environmentally-degradable polymers in plastics waste management

The aim of plastics waste management is to maintain the value of the polymers as long as possible and to keep them within a well-defined network for recycling. Non-degradable polymers in large volumes are essential for the execution of a successful recycling program. In contrast, the resistance to degradation of these materials is not so desirable in products which have a high probability of ending up as litter. Thus manufacturers of biodegradable resins target applications such as degradable commercial fishing nets, biodegradable diapers and plastic bags.

The development of biodegradable polymers is incompatible with collection and recycling incentives. The mixing of inert and biodegradable polymers destroys the performance value of the material and prevents recycling. It is thus crucial that applications for degradables remain within a well-defined network separate from the recycling of traditional plastics.

Beyond the specific applications fields mentioned above, biodegradation is an extravagance when only finite resources are available. It neither conserves the material as mechanical recycling does, nor does it recover any of the energy contained in the material, as does energy recovery.

3.4 APPLICATIONS FOR RECYCLED PLASTICS

The economic driving forces for recycled plastics have been anything but constant during the last 50 years due to large fluctuations in virgin prices and supply. Early recycling efforts were seen only as a means of lowering material consumption within industry by mixing clean, uncontaminated processing waste with virgin material. During the 1960's, virgin material prices fell abruptly and recycling activities stagnated. In the 1970's they picked up again as an oil embargo caused material shortages. During the late 1980's, again, material prices decreased. As legislative pressure on industry and consumer preferences are evolving, however, recycling is becoming less of an option and more of a necessity and business opportunity.

3.4.1   FACTORS AFFECTING MARKET ACCEPTANCE

Successful market penetration depends on an interplay between several critical factors. According to a study made at the University of Stirling, Scotland [78] the three major factors which could boost post consumer polymer waste use in industry are, in order of importance:

- reliable quality,
- comparable price to virgin material,
- reliable supply.

Further factors that will have a severe effect upon the viability of recycling are legislation and consumer attitudes.

Quality

Recycled post consumer material will not be fully accepted in the market until it has attained a quality close to that of virgin material. Nevertheless, its performance will always be inferior to that of comparable virgin material.

Price

Price is a decisive factor for market acceptance. The cost of collection, sorting, and reprocessing constitute the major part of the market price of recycled plastics. Given comparable quality and no external intervention, such as legislation or consumer preferences, recycled plastics will have to undercut the prices of virgin material to be competitive. This is not the case at present although there are exceptions such as recycled PET [18]. Virgin material price fluctuations can quickly eliminate profit margins achieved by stringent management of recycling infrastructure and technology. Annual market price fluctuations of up to 80% can sometimes be seen for virgin plastics. It is often argued that virgin prices do not reflect the environmental costs of extracting raw material. There is currently no way of representing the environmental advantages of recycling in monetary terms.

Supply

A sufficiently large supply of recycled material which does not fluctuate over time is crucial for the development of markets. Shortages can either be due to lack of processing capacity or to low, unstable recovery of plastic waste. There is a hen-and-the-egg problem: industry hesitates to build up processing capacity due to high capital costs and fear for insufficient material supply whereas on the other hand, the storage of recyclables without available reprocessing capacity occupies costly storage capacity.

Consumer attitudes

Consumer attitudes, as reflected by consumer preferences and legislation, play an important role in the viability of recycling. Consumers show increasing acceptance of recycled products, often offered at higher prices and with lower quality than their virgin material alternatives. Consumer organisations have made recycling into a sales advantage by using eco-points or similar classification instruments to determine the cost-environmental performance relationship of products. Nevertheless, in the long run higher quality must be ensured for recycled post-consumer waste to maintain and develop markets.

Legislation

Legislation can strongly affect the business climate by regulating waste management alternatives such as landfill and incineration, as well as by stipulating recovery and recycle goals for individual industries. Current legislation and directives concerning recycling and material recovery place strong emphasis on the collection of material, with few indications as to how markets for these materials will be developed.

An illustrative example of this was the introduction of the German DSD-programme [79, 80]: German authorities failed to take national recycling capacity into account when establishing directives on plastics waste recovery. Nearly twice the amount of available capacity would have been needed to manage the waste which was collected. Furthermore, the sales opportunities for reprocessed plastics were generally 20-25% below the production capacity [81]. The result was that companies did not increase their production even though it was possible. Instead they were even paying to ship their collected waste to other countries.

Perhaps the most persuasive laws are those forbidding or sharply raising the costs of landfill. This provides a strong incentive for companies to re-route their waste into new products rather than into landfills or incinerators. The technological capability to recover and reprocess plastics waste exists within Europe, but economic viability has still not been attained and is not predicted to do so before the end of this century [20].

Packaging waste is under environmental pressure in Europe. The European Council adopted in 1994 the European Parliament and Council Directive on packaging and packaging waste with the following features [82]:

- prevention and reduction of the quantities of packaging waste produced and of the harmfulness of packaging waste;
- promotion of the reuse of packaging;
- recovery of the packaging waste whose generation cannot be avoided;
- reduction of final disposal to the very strict minimum.

The object of the Community policy on packaging is to regulate packaging, but also to streamline national policies on packaging waste. The main issues under discussion are that:

- no later than 10 years from the date by which the Directive is passed, member states must have implemented it in national law; 90% by weight of packaging waste output will be removed from the waste stream for the purpose of recovery, out of which 60% by weight shall be removed for the purpose of recycling;
- final disposal of packaging waste will be limited to no more than 10%;
- "recovery" includes energy recovery.

A further interesting point of the Directive is that some essential requirements concerning the composition of packaging, its reuse, and recycling with respect to the environment are defined as criteria for free trade. This puts high pressure on the packaging industry to better manage their waste streams, minimise the utilisation of packaging, monitor the materials used and provide for efficient recovery and recycling at the end of the service life. The inclusion of incineration as a recovery alternative is a logical and necessary step from an environmental point of view, since multilayer packaging in many cases consumes equal amounts of energy to recycle as is saved it terms of natural resources.

3.4.2   EMERGING MARKETS FOR RECYCLED PLASTICS

Demand for plastics containing recycled material is increasing within several industrial sectors such as automotive, business machines, and electronics.

Automotive industry

Ford has adopted a global target of a minimum of 25% recycle-content for plastic parts to satisfy recycling concerns [35]. Since the introduction of this target, Ford has seen a doubling of post-consumer recyclate use to an annual 11,800 tonnes. An example of secondary applications is Hoechst PET recycle-content grades used in grille-opening retainers. Such a tendency would not have been possible had the plastics producers not responded to demand by developing grades that meet price and performance requirements from the automotive industry. These materials are nevertheless intended for applications where the intrinsic properties of the material are more important than surface appearance and where processing requirements are less stringent.

DuPont recovers used nylon carpets and converts them into new raw material either by mechanical or chemical recycling. The mechanically recycled material has already been approved for high-performance applications such as fan shrouds and other under-the-bonnet automotive applications. The chemical process generates virgin-quality material with up to 50% recycle content to give a new carpet fibre with the same wear resistance, stain resistance, and dyeability as the virgin material.

IT Equipment and consumer electronics

In the United States state authorities are following federal guidelines in demanding material with at least 25% certified post-consumer material for computer housing. OEMs (original equipment manufacturers) are thus looking for suppliers able to deliver such material even at prices well over those of virgin material. Although it is difficult to keep up material quality to the standards of virgin material, recycled material evidently brings added value to the product in this case.

Food packaging

Another large volume market is opening up as the American Federal Drug Agency has approved recycled PET for food contact in food and beverage applications [83]. In response to this, material has been introduced with 25-35% recyclate content in an isolated inner layer for food packaging [84]. An example of such an application in an injected preform for a PET drink bottle is displayed in Figure 3.14 [18]. A layer of 50% recycle-content PET, coloured dark in this case to allow detection, is encapsulated by interior and exterior layers of virgin PET.

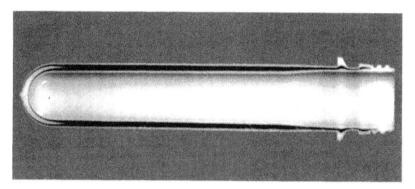

Figure 3.14   PET beverage bottle preform with an interior layer of recycled PET.

Wood replacements and interior design

Recycled mixed plastics are often used as replacement for wood in applications where in-service resistance to mould growth is of primary importance. Furthermore, they can be used as replacement for wood in transport pallets and crates to reduce transport and maintenance costs and transport packaging losses [85, 86]. Since plastics absorb insignificant amounts of moisture compared to wood, weight savings in transport can be considerable. One of the more interesting new application fields is in interior design, where these materials offer design freedom and can be appreciated for what are usually considered to be defects, such as swirling colours and surface roughness which can be used as design features [87].

3.4.3   RESPONSIBILITY IN THE RECYCLING CHAIN

Legislation is increasingly shifting the responsibility for recycling from municipal authorities to manufacturers and distributors according to the "Polluter Pays Principle". This is seen in the laws being passed on the recycling of electronic and automotive waste in several European countries. OEMs and distributors are obliged to take back their products at the end of their service lives, dismantle them, recycle what can be recycled, and dispose of the rest with the least possible effect on the environment. In cases like Germany and The Netherlands financing is provided by industry itself, while in other countries customers pay through fees either upon purchase of disposal [88]. Similarly, packaging legislation singles out private sectors by setting stringent recycling targets [89].

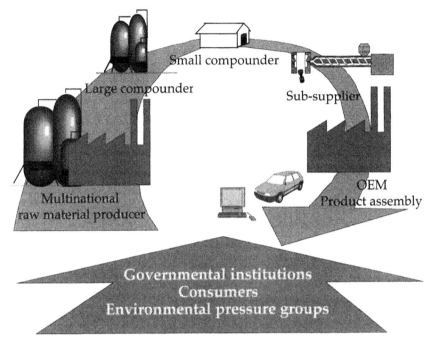

Figure 3.15   Direct pressure for environmental improvement within polymer related industry lies mainly on multinational raw material producers and on OEMs.

Within the production chain it is important to determine which partner assumes responsibility for waste (Figure 3.15). OEMs are already assuming or being given the responsibility by legislation to establish a functioning recycling network. It is less likely that they will take an active part in this network, however; they will rather use recycled material in their products and use their skills in design to facilitate the recovery of material from their products. There are already many recycling and collection networks in action, and there is no reason why an OEM would step away from its core activity to participate in activities better known by others. A whole new material handling industry is emerging to handle the issues of collection, sorting, and cleaning. Raw material producers participate in recycling by providing processing know-how, advice on design for recycling and by co-ordinating smaller material streams into stable supplies of quality recycled material.

Close co-operation between suppliers, OEMs and customers is vital to pin-point secondary applications. In many cases innovation in suitable applications for recycled materials comes from shrewd compounders and reclaimers.

REFERENCES

1. *Plastics Recovery in Perspective: Plastics Consumption and Recovery in Western Europe 1994*, Association of Plastics Manufacturers in Europe (APME), Brussels, Belgium (1996).

2. H. F. Lund Ed., *The McGraw-Hill Recycling Handbook*, McGraw-Hill, Inc.: New York, USA (1993).

3. A. J. Poll, *Recovery Rates Achieved by Kerbside Collection Schemes and their Impact on Residue Disposal* in proceedings of *Waste: Handling, Processing and Recycling*, London, Institution of mechanical Engineers, pp. 53-61 (1993).

4. R. G. Saba and W. E. Pearson, *Curbside Recycling Infrastructure: A pragmatic approach* in *Plastics, Rubber and Paper Recycling: A pragmatic approach*, C. P. Rader et al. Eds., American Chemical Society: Washington D.C., pp. 11-26 (1995).

5. J. P. Trotignon, M. Piperaud, J.Verdu, and A. Dobraczynski, *Précis de Matières Plastiques: Structures-propriétés mise en oeuvre et normalisation*, 5th ed., Pollina Luçon (1993).

6. M. S. Super, R. M. Enick, and E. Beckman, in proceedings of *Recycle '92*, Davos, Switzerland, Maack Business Services (1992).

7. D. Schudel and W. Koller, *Economic Solutions to Materials Recycling of Plastics Using the Buhler NIR Spectrometer NIRIKS* in proceedings of *R'95*, Geneva, Switzerland, **3**, pp. 257-265 (1995).

8. N. Eisenreich, J. Herz, H. Kull, and W. Mayer, *Fast On-Line Identification of Plastics Materials by Near Infrared Sprectoscopy* in proceedings of *R'95*, Geneva, Switzerland, **3**, pp. 230-235 (1995).

9. G. Tersac, M. Rakib, M. Stambouli, J. Deuerlein, and G. Durand, *Purification, par Dissolution Selective, de Déchets d'Objets Usagés Broyés en Polyethyleneterephtalate, Pollués Notamment par du Polychlorure de Vinyle*, European Polymer Journal, **30**, 2, pp. 221-229 (1994).

10. C. D. Papaspyrides et al., *Recycling of Glass Fiber Reinforced Thermo-Plastic Composites*, Resources, Conservation and Recycling, 14, pp. 91-101 (1995).

11. C. D. Papaspyrides et al., *A Model Recycling Process for Low Density Polyethylene*, Resources, Conservation and Recycling, 12, pp. 177-184 (1994).

12. J. G. Poulakis and C.D. Papaspyrides, *The Dissolution/Reprecipitation Technique Applied on High-Density Polyethylene: I. Model Recycling Experiments*, Advances in Polymer Technology, **14**, 3, pp. 237-242 (1995).

13. P. Mapleston, *Auto Parts are Dissolved in New System for Recycling*, Modern Plastics International, **24**, 8, pp. 20-21 (1994).

14. B. K. Mikofalvy and H. K. Boo, *Technical Aspects of Vinyl Recycling in Emerging Technologies* in *Plastics Recycling*, G. D. Andrews and P.M. Subramanian Eds., American Chemical Society: Washington DC, USA, pp. 296-309 (1992).

15. P. Mapleston, Modern Plastics, **69**, 13, p. 34 (1992).

16. C. P. Lambert, *Identification and Separation of Plastics in Mixed Waste* in proceedings of *R'95*, Geneva, Switzerland, **3**, pp. 240-249 (1995).

17. R. Herzog, *PET-Recycling Switzerland: Economic efficiency of a business-led recycling network* in proceedings of *R'95*, Geneva, Switzerland, EMPA, **3**, pp. 180-184 (1995).

18. A. Barrage, LPM AG., Tannenweg 10, CH-5712 Beinwil am See, Switzerland: Personal Communication.

19. A. Stassinopoulos, *A System for Recovery of Materials from Motor Oil Containers with Favorable Economic and Ecological Balance* in proceedings of *R'95*, Geneva, Switzerland, **3**, pp. 146-150 (1995).

20. *Plastics Recovery in Perspective: Plastics Consumption and Recovery in Western Europe 1993*, Association of Plastics Manufacturers in Europe (APME), Brussels, Belgium (1995).

21. V. Williams, *Plastic Packaging for Food: The Ideal Solution for Consumer Industry and the Environment* in proceedings of *R'95*, Geneva, Switzerland, **2**, pp. 49-56 (1995).

22. J. R. Ellis, *Polymer Recycling: Economic realities* in *Plastics, Rubber and Paper Recycling: A pragmatic approach*, C.P. Rader and et al. Eds., American Chemical Society, pp. 62-69 (1995).

23. *Polymer Recovery*, GE Plastics / Ravago Plastics (1991).

24. P. L. Hauck and R. A. Smith, *Integrated Waste Management* in proceedings of *Dornbirn '95* (1995).

25. R. Rieß, *Recycling of Engineering Plastics - Options and Limitations* in proceedings of *R'95*, Geneva, Switzerland, **2**, pp. 24-29 (1995).

26. P. A. Eriksson, *Mechanical Recycling of Glass Fibre Reinforced Polyamide 66*, Departement of Polymer Technology, Royal Institute of Technology, Stockholm, Sweden (1997).

27. J. A. N. Scott, *The Potential for Mixed Plastics Recyclate* in proceedings of *R'95*, Geneva, **3**, pp. 136-141 (1995).

28. E. Klobbie, USA, Patent Nr: US 4,187,352 (1974).

29. *Recycloplast Procedure*, Resource Recycling, **6**, 3, p. 16 (1987).

30. NewPlast Process, NewPlast Process Holding N.V.: ICC, 20, rte de Pré-Bois, 1215 Geneva 15, Switzerland.

31. K. Khait, *Advanced Solid-State Shear Extrusion Pulverization Process for Recycling Unsorted Plastic Waste* in proceedings of *GLOBEC '96, 9th Global Environment Technology Congress*, Davos, Switzerland, pp. 18.3.1-18.3.5 (1996).

32. E. Culp, *Additives Can Give Reclaim a Second Life*, Modern Plastics International, **24**, 9, pp. 59-62 (1994).

33. L. A. Utracki, *Polymer Alloys and Blends: Thermodynamics and Rheology*, Carl Hanser Verlag: Munich (1989).

34. D. Adamus, Press Release Nr 46-94, GE Plastics, 6/6/94 (1994).

35. *Auto Makers Reconcile Quality, Cost and Recyclate Concerns*, Modern Plastics International, **25**, 8, pp. 51-55 (1995).

36. C. B. Bucknall, *Toughened Plastics*, Applied Science Publishers Ltd.: London, p. 20 (1977).

37. S. Rostami, *Polymer-Polymer Blends in Multicomponent Polymer Systems*, I. S. Miles and S. Rostami Eds., Longman Scientific & Technical: Essex, UK, pp. 63-102 (1992).

38. K. E. Van Ness and T. J. Nosker, *Commingled Plastics* in *Plastics Recycling: Products and Processes*, R. J. Ehrig Ed., Carl Hanser Verlag: Munich, pp. 187-229 (1992).

39. R. H. Burnett and G. A. Baum, *Engineering Thermoplastics* in *Plastics Recycling: Products and Processes*, R. J. Ehrig Ed., Carl Hanser Verlag: Munich, pp. 151-168 (1992).

40. F. Maspero, *Etude du Recyclage d'un Film Multicouche: Caractérisation et Propriétés*, Ecole Polytechnique Fédérale de Lausanne, Laboratoire de Technologie des Composites et Polymères, Lausanne, Switzerland, February 1996 (1996).

41. M.-A. Grumser, *Recyclage d'un Film Multicouche à Application Pharmaceutique*, Ecole Polytechnique Fédérale de Lausanne, Laboratoire de Technologie des Composites et Polymères, Lausanne, Switzerland, June 1996 (1996).

42. Y. Wyser, *Life Cycle Study of Multilayer Polymer Films for Packaging Applications* in proceedings of *R'97*, Geneva, Switzerland, **2**, pp. 149-154 (1997).

43. T. P. La Mantia, Associazione Italiana di Scienza e Technologia Delle Macromolecule, **XV**, 36, p. 61 (1990).

44. S. Fuzessery, in proceedings of *Recycle '89*, Davos, Switzerland (1989).

45. A. Boldizar, *Simulated Recycling-Repeated Processing and Ageing of LDPE* in proceedings of *R'95*, Geneva, Switzerland, **3**, pp. 10-18 (1995).

46. *Organic Peroxides Face Stiff Technological Competitors*, Modern Plastics International, **26**, 1, pp. 21-22 (1996).

47. *Organic Peroxides: Viscosity modification of recyclate is one focus of R&D*, Modern Plastics International, **23**, 9, pp. 48-53 (1993).

48. *Blowing Agents: Products minimize tradeoffs as CFC phase-out takes effect*, Modern Plastics International, **23**, 9, pp. 35-37 (1993).

49. *Colorants: Heavy-metal restrictions promote organic replacements*, Modern Plastics International, **22**, 9, pp. 52-53 (1992).

50. *Flame Retardants: Processors learn to work with halogen-free systems*, Modern Plastics International, **23**, 9, pp. 39-41 (1993).

51. W. J. Farrissey, *Thermosets* in *Plastics Recycling: Products and Processes*, R. J. Ehrig Ed., Carl Hanser Verlag: Munich, pp. 233-262 (1992).

52. G. Mackey, *A Review of Advanced Recycling Technology* in *Plastics, Rubber and Paper Recycling: A pragmatic approach*, C. P. Rader and et al. Eds., American Chemical Society: Washington D.C., pp. 161-169 (1995).

53. *Coke and Pepsi will use Recycled PET in Bottles*, Modern Plastics, **21**, 1, p. 45 (1991).

54. K. H. Redepenning, *Waste to Energy With the VTA Pyrolysis and Gasification Processes* in proceedings of *R'95*, Geneva, Switzerland, **5**, pp. 84-91 (1995).

55. P. Groenewegen and F. den Hond, *Technological Innovation in the Plastics Industry and its Influence on the Environmental Problems of Plastics Waste (SAST Project N° 7):*

Report on the Automotive Industry, Commission of the European Communities, EUR-14733-EN, July (1992).

56. W. Gianapini and E. Jacia, *Technological Innovation in the Plastics Industry and its Influence on the Environmental Problems of Plastics Waste (SAST Project N° 7): Report on Substitution of Virgin Plastics by Recycled Material*, Commission of the European Communities, EUR-14735-EN, Sept. (1992).

57. J.-M. Bemtgen, *Integrated Strategy Product Design-Energy -Waste of the EU* in proceedings of *GLOBEC '96, 9th Global Environment Technology Congress*, Davos, Switzerland, pp. 2.2.1-2.2.5 (1996).

58. F. E. Mark, *Energy Recovery through Co-combustion of Mixed Waste Plastics and Municipal Solid Waste* in proceedings of *R'95*, Geneva, Switzerland, **5**, pp. 144-154 (1995).

59. J. Brandrup, Hoechst High Chem Magazine, **10**, p. 36.

60. A. Barrage, *Recovery of Economical Value from Plastic Waste in Switzerland* in proceedings of *ReC '93*, Geneva, Switzerland, **3**, pp. 326-331 (1993).

61. P. Nüesch, *Incineration Problems and Energy Recovery from Plastics Waste* in proceedings of *ReC '93*, Geneva, **3**, pp. 369-399 (1993).

62. J.R. Serumgard and A.L. Eastman, *Scrap Tire Recycling: Regulatory and Market Development Progress* in *Plastics, Rubber and Paper Recycling: A pragmatic approach*, C. P. Rader and et al. Eds., American Chemical Society: Washington D.C., pp. 237-244 (1995).

63. M. Tellerbach, *Waste Conversion to Energy in Cement Plants: A sensible recovery solution, or a seriousthreat to other waste disposal facilities?* in proceedings of *GLOBEC '96*, 9th Global Environment Technology Congress, Davos, Switzerland, pp. 3.1.2-3.1.5 (1996).

64. *Plastics Packaging: Friend or Enemy*, Association for Plastics Manufacturers in Europe (APME), Brussels, Belgium.

65. A. A. Jean, *Co-Combustion: an Efficient and Safe Process for Energy Recovery from Shredding Refuses* in proceedings of *R'95*, Geneva, Switzerland, **5**, pp. 34-40 (1995).

66. S. Dalager, *Energy Recovery Versus Material Recycling, the Danish Situation* in proceedings of *Plastics Recycling '91*, Copenhagen, Denmark, Society of Plastics Engineers, pp. 4.1-4.5 (1991).

67. *Biodegradable Polymers and Recycling* in proceedings of *International Workshop on Controlled Life-Cycle of Polymeric Materials*, Stockholm, Sweden (1994).

68. G. Swift, *Opportunities for Environmentally Degradable Polymers*, J. Macromol. Sc.-Pure Appl. Chem., **A32**, 4, pp. 641-651 (1995).

69. *OS-Order till Plastinject*, Plastforum Scandinavia, 12, p. 22 (1993).

70. G. J. L. Griffin Ed., *Chemistry and technology of Biodegradable Polymers*, Blackie Academic & Professional: Glasgow, UK (1994).

71. P. Krusell, *Säckar och Bestick för Komposten: Starkare Nedbrytbar Plast ska ta Marknadsandelar*, Plastforum Scandinavia, 12, p. 43 (1993).

72. Y. Niyeda, *Polylactic acid, LACEA Development - Fundamental Strategy and Current Strategy* in proceedings of *GLOBEC '96, 9th Global Environment Technology Congress*, Davos, Switzerland (1996).

73. Showa Highpolymer Co., Personal Communication.

74. *Degradable Resin Makers Seek Markets in Japan*, Modern Plastics International, **23**, 10, pp. 26-28 (1993).

75. K. J. Seal, *Test Methods and Standards for Biodegradable standards in Chemistry and Technology of Biodegradable Polymers*, G. J. L. Griffin Ed., Blackie Academic & Professional: Glasgow, UK, pp. 116-134 (1994).

76. ASTM Standards on Environmentally Degradable Plastics, Nr 003-420093-19 (1993).

77. A. C. Albertsson, *Degradable Polymers*, J. Macromol. Sc. - Pure Appl. Chem., **A30**, p. 757 (1993).

78. R. Ball and A. Unsworth, *Markets for Consumer Plastics Waste* in proceedings of *R'95*, Geneva, Switzerland, **1**, pp. 281-286 (1995).

79. K. O. Tiltmann, *Das "Duale System"* in *Reycling Betriblicher Abfälle*, WEKA Fachverlag für Technische Fürungskräfte GmbH: Augsburg, Germany, pp. 11/3.3.1-11/3.3.20 (1992).

80. R. Ball and A. Unsworth, *Markets for Post Consumer Plastics Waste in proceedings* of *R'95*, Geneva, Switzerland, **1**, pp. 281-286 (1995).

81. A. Christensen, *Recycling of Plastics. How will the Political Aims be Reached? The Challenge of the Market* in proceedings of *Plastics Recycling '91*, Copenhagen, Denmark, Society of Plastics Engineers, pp. 5.1-5.6 (1991).

82. European Parliament and Council Directive, Nr 94/62/EC, Art. EC, 31 December 1994.

83. *U.S. FDA OKs Recycled PET for Food Contact*, Modern Plastics International, June, p. 8 (1993).

84. *Precision Degradables Redefines Application Rules*, Modern Plastics International, **25**, 8, pp. 55-57 (1995).

85. L. Martinello, *WoodPack Crates & Technology* in proceedings of *GLOBEC '96, 9th Global Environment Technology Congress*, Davos, Switzerland, pp. 14.8.1-14.8.13 (1996).

86. K. Gustafsson, *Reduce, Reuse, Recycle - Opportunities for Returnable Packaging* in proceedings of *GLOBEC '96, 9th Global Environment Technology Congress*, Davos, Switzerland, pp. 16.2.1-16.2.8 (1996).

87. P. Mapleston, *Recyclate is Breaking Ground in a New Area: Interior Design*, Modern Plastics International, **23**, 6, p. 54 (1995).

88. H. Valentine, *End of Life Electronic Waste Legislation in Europe: Issues and Prospects* in proceedings of *R'95*, Geneva, **1**, pp. 50-56 (1995).

89. G. Vogel, *Efficiency of the Instruments of the Austrian Packaging Ordinance in proceedings* of *R'95*, Geneva, **1**, pp. 69-76 (1995).

4

LIFE CYCLE ASSESSMENT

Life Cycle Assessment is a tool permitting the study and evaluation of emissions and environmental impacts over the entire life cycle of a product or process. The aim is to relate the environmental load to the product functional unit to assess technological improvements for reduced resource use and environmental impacts on humans and on the ecosystem. The use of Life Cycle Assessment (LCA) is increasing in all sectors of society. It is becoming a key component in the monitoring of environmental performance of products and processes as well as an element of product development and long-term strategic planning. Financial institutions are increasingly using environmental performance indicators as criteria for access to finance and insurance policies.

This chapter presents a brief review of the concept of Life Cycle Assessment and its role as a decision-making tool for the plastics industry. The "SETAC LCA-Code of Practice" (1993) is followed with LCA divided into four parts, namely: goal definition and scoping, inventory analysis, impact assessment, and improvement analysis. The inventory analysis is reviewed according to the Centre of Environmental Science of the University of Leiden's guidelines on Environmental Life Cycle Assessment of Products. Special attention is paid to issues specific to polymer, with reference to the inventory analysis methodology proposed by the Association of Plastics Manufacturers in Europe: feedstock energy issues, open- and closed loop recycling and reuse are treated. Guidelines to aid in the analysis of LCA results are given.

4.1 INTRODUCTION

The evaluation of the environmental impact of a product or process requires a means of attributing resource use, waste and pollution. It is often far from easy to decide what is relevant to the product or process under consideration. It can even be difficult to define adequately what the product or process life cycle really is. Earlier, pollution could easily be traced to local sources. Now, as humanity is becoming more mobile, pollution originates in a large number of non-point sources, such as transport and consumption. To fully cover all relevant routes of pollution and energy consumption it is necessary to go beyond the gates of production plants to include flows and transports of raw materials, goods, energy and waste. This is a suitable scenario for Life Cycle Assessment (LCA) and the reason for its rising importance as a tool for performance improvements.

LCA is an analysis system which generates information on the environmental impacts and resource demands of a product or material over its entire life cycle, from the extraction of the natural resources providing materials and energy to the disposal of the object once it has ceased to be useful. The commonly-accepted methodology of LCA is structured in four parts: goal definition and scoping, inventory analysis, impact assessment, and improvement analysis. Figure 4.1 shows this structure. The goal definition and scoping step determines the purpose of the study, the scope, the functional unit and procedures for assuring the quality of the study. An inventory analysis establishes a Life Cycle Inventory (LCI) which quantifies resource and energy consumption and emissions of processes relevant to the functional unit. This step is followed by an impact assessment where the consumed resources and energy and the produced emissions are interpreted to determine their effect on humans and the environment. The improvement analysis is used in the overall evaluation of related aspects, leading to the application of the conclusions in a specific field, be it product or process innovation or achievement of legal compliance.

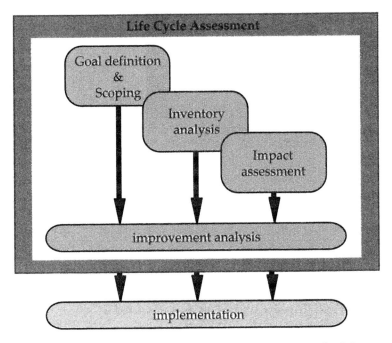

Figure 4.1   The elements of the life cycle assessment methodology.

Since the early 1960's, LCA has been under constant development. Early driving forces were the need for improved methods for product comparison and the energy crisis and solid waste disposal problems. In the late 1980's, LCA was recognised as a potentially useful tool for product analysis and improvement. It is now assuming a dominant role as a support for decision making in several sectors. Industry, public authorities, consumer organisations, insurance companies and banks are frequent users of LCA. Standards have arisen within different business organisations [1-3] and work is well underway towards regional [4] and international standardisation within the ISO network [5].

Performing product-oriented LCAs may become a key to the market place. Consumers are showing an increasing need for information concerning the environmental performance of products and services. It is in the interest of public and private institutions to provide this information. LCAs are already applied to a significant extent as a base for eco-labelling, to guide consumers in product selection. To be able to demonstrate the environmental performance of a company can also be a key to finance, as banks are increasingly using environmental performance indicators as a means of judging long-term profitability and for investment risk analysis.

Industry is rapidly gaining experience in the application of LCA. The predominant motivation for its use in industry is analysis of ones own product to determine environmental- and efficiency weak-points (Figure 4.2). Later additions to the LCA concept are substance flow analysis, which clarifies at what geographical level impacts are active and possible interaction between different types of emissions, and risk assessment, a statistical approach to what impacts are most urgent to prevent, what alternatives are within the financial scope of the company and give the highest cost savings. Indeed, LCA is as much a cost-saving tool as it is a method for lowering environmental impact. In a surprisingly large number of cases, the reduction of environmental

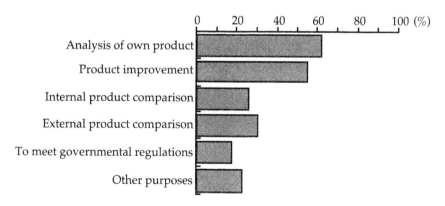

Figure 4.2   LCA applications by frequency of use (after Ryding [6]).

impact translates into cost savings in terms of lower waste generation during production, better material utilisation, and improved process efficiency. Furthermore, the integration of LCA in product development encourages interdisciplinary co-operation to prevent environmental impacts already at the development stage as well as surveillance of impending legislative changes concerning company activities.

LCA studies performed in the past show large inconsistencies, to the point of drawing completely different conclusions from an identical collection of alternatives [7]. These differences may arise from the methodology used or from implicit decisions and assumptions made during the study. There is strong demand for standardisation of the LCA methodology. A leading international organisation in the co-ordination of these efforts is the Society of Environmental Chemistry and Toxicology (SETAC). In this chapter we will follow the methodology proposed in the LCA Code of Practice from the SETAC workshop in Seisimbra, Portugal, 1993 [7]. Illustrative examples are given from a fictional study of an automotive part made from HDPE.

4.2 GOAL DEFINITION AND SCOPING

Goal definition and scoping is the initial step of a LCA. It determines the *purpose of the study, the scope, the functional unit* and a procedure for *quality assurance of the result* (Figure 4.3). Sources for the data to be entered into the analysis should be identified, as should the type of impacts to be evaluated in the impact assessment.

4.2.1   PURPOSE

The reason for carrying out the LCA and the intended application of the results should be clearly defined.

Possible applications are categorised in different ways by different authors. Figure 4.2 displays one set of applications. The SETAC Code of Practice identifies internal and external use as subgroups:

- *Internal LCAs* are not publicly released. This gives more freedom to exchange information. Confidential data can be used, and conclusions can be put into context, while all concerned within the group have easy access to the information upon which the study is based on.

- *External LCAs* are used for communication to the public, governments, consumers, clients, etc. The aim should be to disclose as much information as possible and its sources to boost the credibility of the study.

As target groups differ widely in scientific background and social function it is recommended that the choice of language and terminology used in the study be given thorough consideration. The methodology used should be disclosed to allow the reader to follow and understand the different stages of the study.

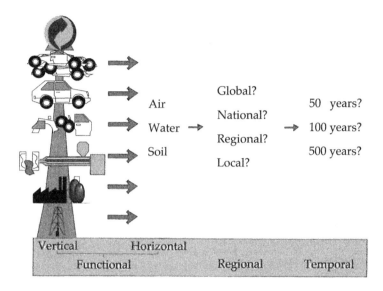

Figure 4.3   Goal definition and scoping determines the object and depth of the study, and the application of the results.

4.2.2 SCOPE

The scope defines the system under consideration and its boundaries. This in turn defines data requirements, necessary assumptions, and the limitations of the study. The scope should be defined to a sufficient breadth and depth for the purpose of the study; his includes *functional, regional*, and *temporal boundaries*.

*   *The functional boundary* has two dimensions: vertical and horizontal. The vertical boundary defines the life cycle stages, that is the processes which are to be included in the study. The horizontal boundary limits the number of impacts to be recorded at each life cycle stage.

*   *The regional boundary* defines the geographical limits within which impacts will be considered (local, national, regional, continental, and global). The definition of these boundaries may depend on the nature of the registered emissions.

*   *The temporal boundary* defines the time period over which impacts will be considered. This depends on the life time of the product, the time horizon for processes, and the longevity of the impact effects on the environmental system.

The appropriate depth of the study can be checked by comparing results from a study including support processes to the main production chain, such as the production of ancillary material, with results from a study excluding them. If the difference between the two is found to be negligible, the boundary can be set to exclude these processes. The excluded processes should be indicated, however, to permit control calculations of external parties. Within the scope it is also important to include estimates on data quality and variability.

Further means of varying the depth of the study have been proposed by Heijungs *et al.* [8]:

- including fewer LCA steps (omitting for example impact assessment, which may be feasible when a product alternative is better than all others in all relevant aspects, or when only a single product is analysed for improvements);
- concentrating on differences between products.

It is clear that limiting the depth of the study has an impact on its reliability especially when certain processes or impacts are omitted.

Fictional HDPE part: Purpose and scope of an LCA.

Purpose: Product improvement

Scope

Functional boundary		Horizontal		
Vertical	Emissions to air	Emissions to water	Solid waste	Energy
Production	CO, CO_2, NO_x, CH_4, SO_x, HCl	$COD^{\wedge\wedge}$, NH_4^+, PO_3^{2-}, H^+	kg	feedstock / fuel
Fabrication	CO, CO_2, NO_x, CH_4	—	kg	feedstock / fuel
Use	CO, CO_2, NO_x, CH_4	—	kg	feedstock / fuel
Disposal	—	—	kg	—
Regional boundary*	g, r, n, l	l	l	l
Temporal boundary**	e	e	e	e

* g = global, r = regional, n = national, l = local ; **e= depending on longevity of emissions

^^ COD = amount of oxidisable material emitted to water

4.2.3 DETERMINING THE FUNCTIONAL UNIT

The functional unit is the basic quantity for the LCA process. In defining the functional unit, it is important to understand the utility of the object under consideration. It can be misleading to take the functional unit to be the production unit. As an example, we will consider an LCA of a car door. At first sight, the door may seem to be the obvious functional unit, being an easily identifiable entity with a clear function. However, its ready utility is not just to look like a door: it contributes to the provision of transport. The most appropriate functional unit would thus be a "car•km", since the door is acquired with the purpose of transporting a certain distance.

A more general definition of the functional unit would thus be the service delivered to a customer or a process by one or several sub-systems, or a measure of the performance which the system delivers. Generally, the functional unit consists of a unit and a quantity (Table 4.1) such as: person-transport kilometres, listening to the radio for one hour, or consuming 200 grams of biscuits.

There are, however, cases where the functional unit does not possess both a unit and a quantity. Heijungs *et al.* [8] take up flowers as a good example of a functional unit that has more to do with a gesture than a quantity. In general a customer asks for a "bunch of flowers", not for "flowers for ten days" or "500 grams of flowers".

Table 4.1   Production unit and functional unit: some examples.

Product	Function	Production unit	Functional unit
paint	protect; decorate	1 kg	m^2•years
shoes	protect; good appearance	1 pair	man•years
car door	permit entry and exit; isolate from ambient environment, noise; protect at impact	1 door	car•km

Many products have several functions, and it can be difficult to determine which one of them is central to the product. A warm sweater, for example, is not only bought to keep a person warm, since its appearance also plays a role in the decision to purchase. The main utility of the sweater being to keep its owner warm, however, appearance can be considered to be secondary in this context. When comparisons between different product alternatives are being made in LCA, they should always be made on the basis of equal functionality. In cases where products provide the same main function but have different subsidiary functions these functions must be clearly defined to allow judgement of their relevance for the final selection.

Once a rigorous functional unit has been defined, the systems delivering the defined functional unit can be evaluated in terms of their environmental efficiency in bringing their service to the user.

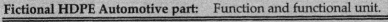

Fictional HDPE Automotive part: Function and functional unit.

PRODUCTION UNIT: One protective shield

FUNCTION: Protect engine, guide air

REQUIREMENT: High toughness-to-weight ratio

FUNCTIONAL UNIT: Car·km

4.2.4 DATA QUALITY ASSESSMENT

The quality of input data is important for the credibility of the study. No matter how performant the methodology, it can never compensate for poor data input. The evaluation of data quality should:

- make as few assumptions and simplifications as possible;
- specify data sources, age and acquisition methods when possible;
- ensure that data on operations represent normal operating conditions;
- verify the reliability of in-house data by comparing to other sources;
- verify the reliability of externally-collected data by comparing several sources.

Special attention should be paid to conditions that are specific to the considered operations. As an example, very few industrial plants operate in isolation: some services, such as the treatment of water effluents and the provision of steam, are often common to several plants within the production site. If the distribution of these services is not well-defined, it can be very tricky to quantify the consumption of each plant or process. Furthermore, if a process produces several useful outputs, the consumption of energy and raw materials of the process must be distributed between these outputs in a rational and transparent manner.

Companies generally only keep the minimum of records necessary to run their plants with respect to processing conditions, legislative requirements and plant operation permits. The information necessary for a LCA may not be available, or it may be available but not in a useful form. For example, if emission figures are available in terms of concentrations it may be difficult to explain them, since it may not be clear at what distance from the source of emissions the concentration has been calculated or under what wind and weather conditions.

4.3 INVENTORY ANALYSIS

In the inventory analysis the product system, is analysed. By this is meant the system of economic and physical processes providing the service defined by the functional unit. The aim is to quantify all the economic and physical inputs and outputs of the chosen product system. This involves quantifying all material and energy flows across the system borders.

The methodology of inventory analysis is well-developed in comparison to the other components of LCA. Some practitioners still prefer to use the inventory as the sole base for improvements, claiming that the impact assessment step is still not sufficiently developed. While it is true that the inventory stage is sufficient to simply identify the emissions, it does not evaluate their relative environmental impacts. For this an impact assessment is necessary.

The SETAC Code of Practice defines five important elements of an inventory analysis:

- definition of system and system boundaries,
- drawing up process flow charts,
- collection of data,
- application of allocation rules,
- treatment of energy.

This is not a chronological order of activities. Heijungs *et al.* [8] propose four steps which describe tasks in a clearer chronological order:

- drawing up the process tree,
- entering the product data,
- applying the allocation rules,
- creating the inventory table.

4.3.1 GENERATING THE PROCESS TREE

The process tree consists of inputs, outputs, waste and emissions of the set of economic and physical processes which contribute to the service provided by the previously-determined functional unit. It is drawn up based on the functional unit itself and the scope of the study. There is agreement at large on applying a "cradle-to-grave" approach, to cover the steps from the extraction of natural resources to waste management, including reuse, recycling, incineration, and disposal.

Once the process tree is complete the delineation of the system boundaries can take place. There are three boundaries to consider:

- the boundary between the product system and the environmental system
- the boundary between relevant and irrelevant processes
- the boundary between the product system and other product systems

To delineate, all economic inputs and outputs need to be traced back to environmental inputs and outputs starting from the functional unit. This creates a natural border between the product system and the environmental system. The chain of processes is only broken when there is recycling to or from other product systems (open-loop recycling). The treatment of emissions in the environmental system is not a part of the product system that produced them: it must be considered as an individual system.

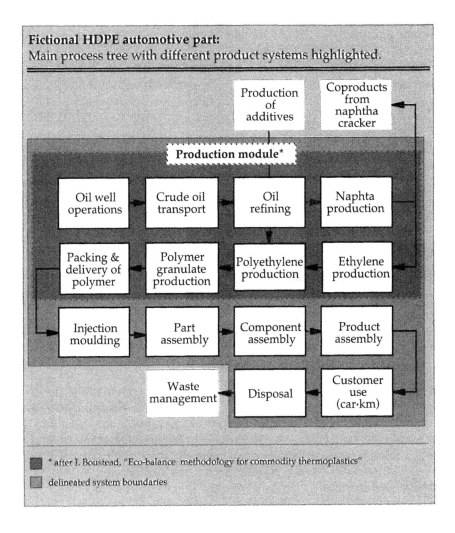

Fictional HDPE automotive part:
Main process tree with different product systems highlighted.

* after I. Boustead, "Eco-balance methodology for commodity thermoplastics"

delineated system boundaries

All processes need energy, raw materials and capital goods such as production tools. To make a plastic bottle it is necessary to have among other things a polymer, energy and a blow moulder. To produce these products requires several other components. If this is continued for all ancillary material and supporting processes the system will grow exponentially and become impossible to handle. In the determination of the boundary between relevant and irrelevant processes no process in which maintenance and depreciation are substantial parts of the product price should be excluded. Sensitivity analysis can indicate whether an excluded process is irrelevant or not. Typically, processes contributing to less than 1% of overall impact are left out.

With polymer products, most processes generate more than one marketable product. In-plant trimmings, runners, and off-specification products can be used as input to other product systems. To avoid an overwhelmingly large product system, most practitioners propose that the process tree be cut at the input of the recycled material. Those processes that are excluded must nevertheless be clearly described so as to preserve transparency.

4.3.2 ENTERING THE PROCESS DATA

In this step data is collected for all the relevant processes in the process tree. In cases where there are no standardised production modules for processes, empirical data has to be identified. This empirical data must be based on a statistically-relevant time period and the averaging and weighing methods must be thoroughly reported.

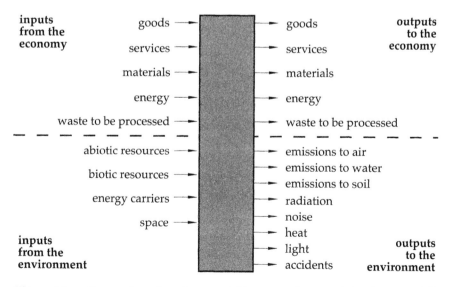

Figure 4.4   Economic and environmental inputs and output (after Heijungs [8]).

Two important aspects of data collection have to be fulfilled to as large extent as possible:

- quantifying the inputs and outputs,
- demonstrating the representativeness and quality of the data.

In the quantification of inputs and outputs Heijungs *et al.* [8] propose a conceptual format for the main structure of data and a technical format which provides the technical rules for entering the data. They distinguish economic and environmental inputs and outputs to each process according to Figure 4.4.

4.3.3   APPLYING THE ALLOCATION RULES

In cases where processes cannot be defined to the most elementary level, calculations have to be made before entering the process data and aggregating it to the inventory table. These calculations are called *allocation* or *partitioning* and consist of dividing inputs and outputs of a *multiple output process* among the co-products according to a set of pre-defined *allocation rules*. The most frequently-used rule is allocation according to mass. Allocation according to economic value is also frequently used, for example in the co-production of steam and electricity. Since economic values are affected by factors such as supply and demand, politics and regulations, they are generally to be avoided.

The fictive single processes resulting from the allocation process should add up to the original multiple process if their respective inputs and outputs are added together. Some examples of multiple processes are:

- co-production: concurrent production of several materials, products or services with a positive value;
- combined waste recycling: concurrent processing of waste with negative value;
- closed-loop recycling: processing waste from one product system to material that can be used as input to another product system within the same plant;
- open-loop recycling: when recycled material coming from another system is used in the considered process.

A typical example of co-production is the naphtha cracker, displayed in Figure 4.5, which produces ethylene, propylene, butadiene and white spirits outputs from inputs of naphtha and energy. The goal with allocation in this case is to create four apparently individual processes with a specific amount of a single product output from a system with several co-products.

Combined waste processing are processes such as mixed waste incineration and landfill, where outputs in terms of emissions to air, water and soil should be allocated to each of a multitude of materials being fed to the process.

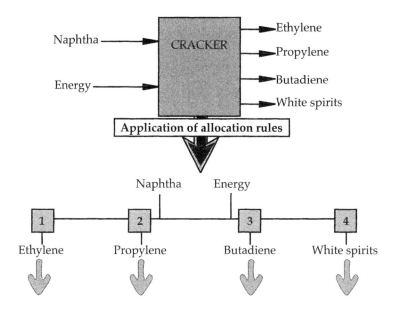

Figure 4.5   Allocation of co-products in a naphtha cracker.

What rule of allocation should be used? The SETAC methodology maintains that "the choice of allocation basis should be related directly to the chemistry and physics of the process". This is one type of *causal allocation*. The analysis itself may consequently be partly chemical-analytical and partly physical. As a principle, this type of allocation should be used wherever possible. In cases where this is not possible *overall apportioned allocation* must be used, based on the functional unit. This means that allocation is made, for example by weight in automotive applications, while for surface treatment it would be by covered surface.

Allocation has been thoroughly discussed in the 1994 SETAC European Workshop on Allocation in LCA [9].

Fictional HDPE automotive part: Allocation assumptions.

- no recycling
- no combined waste recycling
- no coproduction more than that was considered in the production module, that is allocation according to the relative mass of each co-product

4.3.4   CREATING THE INVENTORY TABLE

Once the environmental impacts have been quantified as fully as possible, the resulting list of impacts is called an inventory table. It consists of the

Fictional HDPE automotive part: Inventory table (after Boustead [11]).

Inputs		Unit	Amount	Outputs		Unit	Amount
Production of 1.05 kg HDPE							
Fuels	Coal	MJ	2.300	Air emissions	CO	g	1.000
	Oil	MJ	2.660		CO_2	g	987.0
	Gas	MJ	11.10		SO_x	g	6.000
	Hydro	MJ	0.410		NO_x	g	11.00
	Nuclear	MJ	1.360		CH_4	g	0.053
	Other	MJ	0.010		HCl	g	0.053
	Total	MJ	17.80	Water emissions	COD	g	0.210
Feedstock	Coal	MJ	0.010		NH_4^+	g	0.010
	Oil	MJ	35.24		PO_3^{2-}	g	0.001
	Gas	MJ	32.00		H^+	g	0.100
	Total	MJ	67.25	Solid waste*		g	32.00
Total feedstock/fuel		MJ	85.10	Useful products	HDPE	g	1050
Raw materials	Iron ore	g	0.315				
	Limestone	g	0.210				
	Water	g	9975				
	Bauxite	g	0.210				
	NaCl	g	4.000				
	Clay	g	0.021				
	Ferromanganese	g	0.001				
Fabrication of part							
Fuels	Coal	MJ	3.445	Air emissions	CO	g	0.250
	Oil	MJ	1.230		CO_2	g	730.0
	Gas	MJ	1.365		NO_x	g	2.300
	Hydro	MJ	3.575		CH_4	g	2.000
	Nuclear	MJ	3.367	Solid waste*		g	23.00
	Other	MJ	0.018				
	Total	MJ	13.00				
Feedstock	Oil	MJ	25.59				
	Gas	MJ	23.24				
	Total	MJ	48.82				
Total feedstock/fuel		MJ	61.82				
Use							
Fuel	Gasoline	MJ	230.4	Air emissions	CO	g	122.0
					CO_2	g	14700
					NO_x	g	31.00
					CH_4	g	15.00
				Solid waste*		g	13.00
Disposal							
				Solid waste*		g	1000

* 56% mineral waste, 19% non-toxic chemicals, 16% slag & ash, roughly 9% industrial waste and a very small amount of toxic waste accounted for in terms of Solid Waste Burden (kg).

contributions of each process calculated according to the allocation assumptions. If possible, data should be presented in the form of averages and range, or some other measure of variability around an average.

Non-quantifiable impacts, such as damage to aesthetic values (for example destruction of landscapes by strip mining, fragmentation of natural areas by roads) must also be included as part of the inventory. These impacts are presented as *qualitative aspects*.

4.4 IMPACT ASSESSMENT

Impact assessment is the step that evaluates the effects of the environmental impacts recorded in the inventory table. There is little agreement regarding a standard methodological framework; The SETAC Code of Practice proposes:

- classification,
- characterisation,
- evaluation.

which have been described in detail in the 1994 report from the SETAC workshop "Integrating Impact Assessment into LCA" [10].

4.4.1 CLASSIFICATION

The data from the inventory table is grouped into a number of impact categories. One impact type may be included in several impact categories, for example, NO_x may be included in acidifying and nutrifying categories.

Table 4.2 List of widely-recognised problems that can be investigated with the standard classification model (after Heijungs [8]).

Depletion	Pollution	Damage
• depletion of abiotic resources	• enhancement of the greenhouse effect	• damage to ecosystems and landscapes
• depletion of biotic resources	• depletion of the ozone layer	• victims
	• human toxicity	
	• ecotoxicity	
	• photochemical oxidant formation	
	• acidification	
	• nutrification	
	• waste heat	
	• odour	
	• noise	

A common classification of categories is according to their role in resource depletion, pollution or damage (Table 4.2). Energy production is often presented as Gross Energy Requirement (GER), while waste production is represented by a Solid Waste Burden (SWB).

4.4.2 CHARACTERISATION

After the classification, the impacts must be qualified and aggregated. The amounts of each resource used or pollutant produced, as determined in the inventory table, are weighted to indicate their importance (using an equivalency factor) and added up according to aggregation criteria.

Two common examples of equivalency factors are Ozone Depletion Potential (ODP) and Global Warming Potential (GWP) displayed in Table 4.3. These are theoretical values which are calculated on a relative scale for different substances.

Values for human and ecotoxicity in Table 4.3 neglect the effects of degradation and transport phenomena. Several authors have concluded that the life-time and volume of dilution of a pollutant need to be taken into account since they affect the concentration increase due to an emission [12-14]. Jolliet [14] has proposed a method called "Critical Surface-Time" which compares the concentration increase caused by an emission to the maximum concentration limit. The maximum concentration limit is defined as the level of concentration at which there is no effect on the most sensitive species, be it human, animal, plant or another organism. Pollution up to the no-effect level of the entire soil, volume of drinking water or air is considered equally detrimental. The effect is also considered to be proportional to the area and time duration over which the concentration increase occurs. Impacts are expressed as an equivalent surface polluted up to the concentration limit during one year. The relation between dilution volume and critical surface can be expressed as the following: if the maximum concentration in the air of NO_x is $3 \cdot 10^{-5}$ g / m^3, its life-time $3 \cdot 10^{-3}$ years and its dilution volume 440 m^3 / m^2, $440 \cdot 3 \cdot 10^{-5} / 3 \cdot 10^{-3} = 4.4$ g of emitted NO_x would be able to fill the air above 1 m^2 of ground up to the concentration limit during one year. Consequently, the equivalent polluted soil surface of 1g NO_x would be 0.22 m^2 per year.

The result of the characterisation is the impact or environmental profile, consisting of impact scores obtained by the use of equivalency factors for each aggregation criterion. These scores are completed with the non-quantifiable impacts that were found in the inventory analysis.

Table 4.3   Impact scores, units, and equivalency factors for the characterisation (after Heijungs [8]).

Impact score	Unit	Equivalency factor
abiotic depletion	—	1/reserves
biotic depletion	yr.$^{-1}$	Biotic Depletion Factor (BDF)
greenhouse effect	kg	Global Warming Potential (GWP)
ozone depletion	kg	Ozone Depletion Potential (ODP)
human toxicity	kg	Human Toxicological Classification Factors
		Air (HCA)
		Water (HCW)
		Soil (HCS)
ecotoxicity		Ecotoxicity Classification Factors
aquatic ecotoxicity	m^3	Aquatic ecotoxicity (ECA)
terrestrial ecotoxicity	kg	Terrestrial ecotoxicity (ECT)
oxidant formation	kg	Photochemical Ozone Creation Potential (POCP)
acidification	kg	Acidification Potential (AP)
nutrification	kg	Nutrification Potential (NP)
aquatic heat	MJ	1
malodorous air	m^3	1/ Odour Threshold Value (OTV)
noise	Pa^{2}s	1
damage	m^{2}s	1
victims	—	1
energy use	MJ	Gross Energy Requirement (GER)
solid waste	kg	Solid Waste Burden (SWB)

Fictional HDPE automotive part:
Equivalency factors for selected emissions.

emission	GWP$_{100}$*	ODP	AP	NP	SWB
Air emissions					
CO	2	—	—	—	—
CO$_2$	1	—	—	—	—
SO$_2$	—	—	1	—	—
NOx	40	—	0.7	0.13	—
CH$_4$	11	—	—	—	—
HCl	—	—	0.88	—	—
Water emissions					
COD	—	—	—	0.022	—
NH$_4{}^+$	—	—	—	0.33	—
acid as H+	—	—	0.88	—	—
phosphate	—	—	—	1	—
Solid waste					
mineral waste	—	—	—	—	1
non-toxic chemicals	—	—	—	—	1

* The index 100 of the of the GWP value is the time period during which effects are considered. The estimated life time of CO$_2$ in the atmosphere is approximately 120 years.

Fictional HDPE automotive part:
Impact scores for the production, use, and disposal
of an HDPE automotive part.

emission	GER (MJ)	GWP$_{100}$ (g)	ODP (g)	AP (g)	NP (g)	SWB (g)
Energy requirements						
	377.3					
Air emissions						
CO	—	246.5	—	—	—	—
CO$_2$	—	16417	—	—	—	—
SO$_2$	—	—	—	6	—	—
NO$_x$	—	1772	—	31	5.76	—
CH$_4$	—	188.1	—	—	—	—
HCl	—	—	—	0.047	—	—
Water emissions						
COD	—	—	—	—	0.022	—
NH$_4^+$	—	—	—	—	0.033	—
acid as H+	—	—	—	0.088	—	—
phosphate	—	—	—	—	0.001	—
Solid waste						
	—	—	—	—	—	1068
Environmental profile	377.3	18624	0	37.135	5.861	1068

Note that due to 5% material loss in the production stage, the production stage requires
1.05 kg of HDPE, while the following steps base calculations on 1.0 kg of HDPE.

4.4.3 EVALUATION

The aim of evaluation is to weight the impact scores according to their relative
significance. This is especially important in comparative studies between
different product alternatives.

If, for example, one alternative has a higher score on global warming, and the
other a higher score for human toxicity, then it is not simple to decide which is
the better alternative if there is no priority scale for the importance of each
score. If human toxicity is judged as a more immediate problem then the best
product alternative will be the one with the lowest score in this respect.

This approach only takes the quantitative aspects of impacts into account. Two
types of decision theory techniques can be distinguished:

*   quantitative multicriteria analysis,
*   qualitative multicriteria analysis.

Quantitative multicriteria analysis defines and applies weighting factors to the environmental profile to permit aggregation of the quantitative part of the inventory table into an unambiguous *single score*. This score is then used to rate different product alternatives. The advantages with this method are that the decision process is easily reproducible and that product comparison is simplified. The risk is that the method of aggregation may imply a scientific accuracy that not necessarily exists. Qualitative aspects of the impacts can not be included in the obtained score. These must be evaluated separately.

Qualitative multicriteria analysis is more informal than quantitative analysis. The assessment and rating of different product alternatives is made by an expert panel. The principal advantage of this method is that it is likely to result in a judgement. In addition it allows the inclusion of quantitative aspects in the decision process. However, it is not well suited to comparisons of more than a few products, since the amount of information becomes overwhelming.

Multicriteria analysis must be followed by an analysis of the reliability and validity of the result. The consequences of data uncertainty and of the assumptions and choices made during the study should be evaluated.

4.5 IMPROVEMENT ANALYSIS

Improvement analysis has three purposes (Figure 4.6): to identify environmental weak points in the production system, technology for improving the system's environmental performance and the organisational measures required for such improvements.

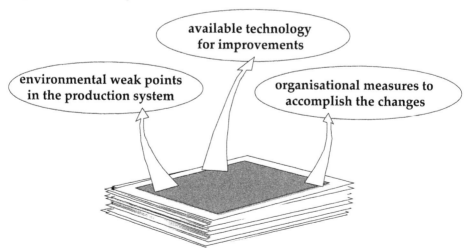

Figure 4.6   The three aims of improvement analysis.

Either the inventory analysis or the environmental profile can be used to identify aspects in need of improvement. Once the processes and substances with the most significant contributions to the environmental profile have been identified, the search for more environmentally-friendly alternatives can start. These alternatives then have to be evaluated individually by experts within, for example marketing, processing and finance and then compared with each other to establish the most feasible alternative.

Fictional HDPE automotive part : Improvement analysis.

In the HDPE automotive case, the inventory table indicates that the environmental impacts of the use of the car dominate during the life cycle of the part. From a polymer processor's point of view the most efficient way of lowering the environmental impact would be to reduce the weight of the part. This could be achieved by various means such as:
- adding short fibre reinforcement to increase the intrinsic stiffness of the material;
- combining short and long fibre reinforced material by integrated processing to achieve optimal weight savings.

The environmental profile obtained for the automotive part indicates that air emissions dominate for the impact scores selected in the example. From the inventory analysis it can be concluded that the dominating emissions (89% by weight) of greenhouse gases emanate from the combustion of carburants during the service life of the car. An overall weight saving of 20% would reduce the impact score for global warming by 18%.

If avoiding solid waste were the highest priority, improvements in process efficiency and recycling of the waste material at the end of the service life would allow the highest reduction of impacts.

4.6 CONCLUSIONS AND POINTERS

There can be large discrepancies in the results of LCAs even when performed on identical product alternatives. Hence, it is important to know how to identify the differences between studies which can explain these discrepancies. The following questions address the most frequent sources of discrepancies:

- What is the functional unit?
- What are the boundaries of the product system?
- What is the basis of the data used in the study?
- According to what methods has the impact assessment been made (try two or three other methods to confirm the results)?
- Is information available on the original recording of emissions to ensure coherence with the functional unit and system boundaries?

It is important to ensure that the chosen functional unit is suitable for the goal of the study. The functional unit is the basis for the definition of the product system and thus also affects the selection of processes and impacts to be included in the study. Using a process tree that is incoherent with the goal or with the functional unit will certainly provide a result, but this result will not necessarily be relevant to the question posed in the beginning of the study.

The relevant sources and geographical and temporal relevance of data, should be checked. If data has been transferred from other regions their relevance to conditions in the region under consideration must be verified. Energy data, for example, can pose a problem since the efficiency with which a unit of energy is generated differs between different regions and does not translate into the same amount of primary energy removed from the environmental system.

As already pointed out, impact assessment methodologies are undergoing intense development, and new approaches regularly appear. It may be worthwhile to try a number of different methods to confirm the result of impact assessment.

It is easy to highlight weaknesses of LCA, but this is not a sufficient argument to abstain from its use. Despite the inherent uncertainties of LCA, it has been shown to be an efficient tool for improving the environmental performance of products and processes.

Complete LCAs are still very time consuming. Periods of up to three person-months are frequent for studies where considerable amounts of data are already available, and times approaching a person-year may be necessary for studies with lower availability of data [8]. The latter kind is currently only justified for long-term strategic planning or compliance with legislation.

There is strong demand for rapid and user-friendly computer-based LCA software and databases. SimaPro 3.1 [15], EPS [3], EcoManager CTM [16], and the Boustead LCI model [17] are examples of software developed to simplify and accelerate the LCA process. Several databases are under development. The Plastics Waste Management Institute (PWMI) is preparing a database with environmental profiles for the production of polymers based on European industrial averages in co-operation with I. Boustead [2, 11, 18-23]. The Society for the Promotion of Life Cycle Development (SPOLD) has collected already existing data from different LCA stakeholders, to provide a publicly available database for Life Cycle Inventories [24]. The number of practitioners of LCA is ever increasing, and the increased availability of LCA data combined with user-friendly computer-based methods will surely establish LCA as an invaluable component of the management tool-box.

REFERENCES

1.   C. Kaniut and H. Kohler, *Life Cycle Assessment (LCA)-A Supporting Tool for Vehicle Design?* in proceedings of *Life Cycle Modelling for Innovative Products and Processes,* Berlin, Germany, Chapman and Hall, pp. 445-457 (1995).

2.   I. Boustead, *Eco-Balance Methodology for Commodity Thermoplastics,* Association of Plastics Manufacturers in Europe, Brussels, Belgium, December (1992).

3.   *The Product Ecology Project: Environmentally-Sound Product Development Based on the EPS System (Environmental Priority Strategies in Product Design),* Federation of Swedish Industries (1993).

4.   L. G. Lindfors, *Nordic Manual on Product Life Cycle Assessment-PLCA,* Conference proceedings, Copenhagen, Denmark (1994).

5.   ISO/TC 207, *ISO 14040-14049-Life Cycle Assessment (LCA), ISO/CD 14040 Life Cycle Assessment: General principles and practices,* Zürich SNV (1994).

6.   S. O. Ryding, *International Experiences of Environmentally-Sound Product Development Based on Life-Cycle Assessment: Final Report,* Swedish Waste research Council, ISSN 1102-6944, ISRN AFR-R-36-SE (1994).

7.   *Guidelines for Life-Cycle Assessment: A "Code of Practice"* in proceedings of *SETAC Workshop,* Sesimbra, Portugal, Society of Environmental Toxicology and Chemistry (SETAC) (1993).

8.   R. Heijungs, *Environmental Life-Cycle Assessment of Products,* Center of Environmental Science, University of Leiden: Leiden (1992).

9.   *Proceedings of the European Workshop on Allocation in LCA,* Center of Environmental Science, Leiden University, Leiden, The Netherlands, SETAC-Europe (1994).

10.   *Integrating Impact Assessment into LCA* in proceedings of *LCA Symposium,* Brussels, Belgium (1994).

11.   I. Boustead, *Eco-Profiles of the European Plastics Industry, Report 3: Polyethylene and Polypropylene,* European Center for Plastics in the Environment, Brussels, Belgium, May (1993).

12.   R. Heijungs and J.B. Guinée, *Using Multimedia Environmental Models in LCA: The flux pulse problem,* CML paper Nr 18 (1994).

13.   D. Mackay, *Multimedia Environmental Models: The fugacity approach,* Lewis Publ., Inc.: Chelsea, UK (1991).

14.   O. Jolliet, *Critical Surface-Time, an Evaluation Method for Life Cycle Impact Assessment Including Fate* in proceedings of *4th SETAC-Europe Congress,* Brussels (1994).

15.   *SimaPro, 3.1,* PRé Product Ecology Consultants, Bergstraat 6, 3811NH Amersfoort, The Netherlands (1991).

16.   *EcoManager,* Franklin Associates Ltd., 4121 W-83rd St., Suite 108, Prairie Village, KS 66208, USA.

17.   I. Boustead, *LCI Model,* Open University of the U.K., St. James House, 150 London Road, East Grinstead, RH191ES West Sussex, UK.

18. I. Boustead, *Eco-Profiles of the European Plastics Industry, Report 2: Olefin Feedstock Sources,* European Center for Plastics in the Environment, Brussels, Belgium, May (1993).

19. I. Boustead, *Eco-Profiles of the European Plastics Industry, Report 4: Polystyrene,* European Center for Plastics in the Environment, Brussels, Belgium, May (1993).

20. I. Boustead, *Eco-Profiles of the European Plastics Industry, Report 5: Co-Product Allocation in Chlorine Plants,* Association of Plastics Manufacturers in Europe, Brussels, Belgium, April (1994).

21. I. Boustead, *Eco-Profiles of the European Plastics Industry, Report 6: Polyvinyl Chloride,* Association of Plastics Manufacturers in Europe, Brussels, Belgium, April (1994).

22. I. Boustead and M. Fawer, *Eco-Profiles of the European Plastics Industry, Report 7: PVDC (Polyvinylidene Chloride),* Association of Plastics Manufacturers in Europe, Brussels, Belgium, December (1994).

23. I. Boustead, *Eco-Profiles of the European Plastics Industry, Report 8: Polyethylene Terephthalate (PET),* Association of Plastics Manufacturers in Europe, Brussels, April (1995).

24. P. Hindle and N. Tieme de Oude, *SPOLD-Society for the Promotion of Life Cycle Development,* The International Journal of Life Cycle Assessment, **1**, 1, pp. 55-56 (1996).

5

LIFE CYCLE ENGINEERING IN PRODUCT DEVELOPMENT

Product and process development is discussed in the light of the Life Cycle Engineering concept. The issues of material reduction, material life extension, product life extension, process improvement and product management are addressed. Examples of products and processes designed with these issues in mind are given. Existing design methodologies already cover design for recycling, assembly and disassembly; in this chapter the importance of including material-related life cycle performance issues, such as durability and reliability, into design strategies is emphasised.

5.1 RESOURCE OPTIMISATION

Current economic processes are resource intensive: merely 5% of the consumed natural resources are transferred into useful products, the remaining 95% constituting byproducts and waste [1]. In such a situation there is a clear need for design practices which address an optimal use of resources. Companies world-wide are moving rapidly towards integrating environmentally-conscious technology and products into their business strategies. The experiences of those which have successfully implemented life-cycle based environmental measures reveal that it encourages innovative product simplification, attracts new customers, and reduces the cost of production and waste management [2].

Life Cycle Engineering can help to identify and reduce the most significant environmental effects of a product. When the idea of environmentally-friendly design was first introduced, there was a tendency to assume that negative environmental impacts could all be designed out of a product by ensuring recyclability [3]. However, experience has shown that product characteristics and patterns of use are determinant for environmental performance, and that the whole life cycle must be considered rather than just recycling. The earlier in the development process this can be recognised, the greater the chances of ensuring significant improvements.

It is important to remember that environmentally conscious design is not an end in itself. A good environmental strategy for a product will satisfy the entire set of product requirements, including performance and development costs. Initial estimations of the readiness of consumers to pay premiums for recyclability and environmental improvements have proved to be exaggerated. Many "green" products have failed on the market by not meeting performance requirements. If environmental considerations are to gain a place in product development they must be accompanied by economic and performance benefits.

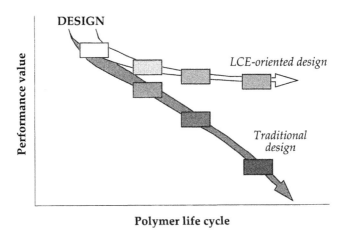

Figure 5.1   LCE aims at maintaining material value throughout the life cycle.

Figure 5.1 is based on the approach introduced in Chapter 1 and shows the potential effect of considering environmental performance in design. The options for improving environmental performance of the value-added chain of a product are shown in Figure 5.2. The aim of LCE is to integrate these options into the product development process.

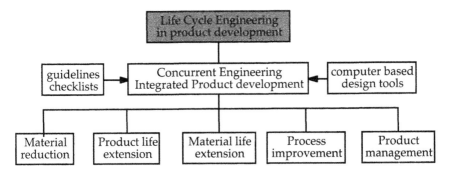

Figure 5.2   Life Cycle Engineering in product development.

The most obvious alternative is to *reduce material use* wherever possible, both in the product itself as well as in ancillary material. The next step would be to extract as much value out of the product and the material of which it is made. This can be achieved by *extending the life of the product* through improved maintenance, reuse, or remanufacturing, and by *extending the life of the material* through revitalisation and/or recycling. Proper material selection for a given application increases the possibility of maintaining material performance value at a high level throughout its life cycle.

Furthermore, *process efficiency* and *product management* can be improved. Process improvement is achieved not only through reduced energy consumption, but also in terms of the use of ancillary equipment and improvements in infrastructure. Management of the environmental effects of products implies regular reviews of product environmental performance as an aid in evaluating product renewal strategies. It also serves to identify products that are particularly hazardous to the environment.

5.2 THE PRODUCT DEVELOPMENT PROCESS

Product development is crucial in determining the final cost and environmental performance of a product. Roughly 70% of the life cycle cost is already defined at the concept design stage [3, 4] (Figure 5.3). There is good reason to believe that the same goes for the environmental effects of a product, since an increasing proportion of life cycle costs is due to environmental legislation,

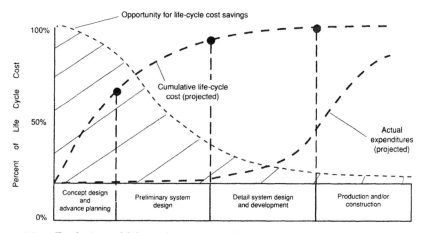

Figure 5.3   Evolution of life cycle cost as a function of product development B. S. Blanchard "System Engineering Management", © 1991, p. 8, Reprinted with permission of John Wiley & Sons Inc. [4].

recovery and recycling costs, or land-fill costs. It is clear that competitive advantage can be gained by taking such considerations into account during product development.

Traditional product development has aimed to manufacture products as efficiently as possible within the frame of an established product specification. Improved production equipment, working methods, and planning have reduced the direct variable labour costs to 10-20% of total manufacturing costs. This approach reaches inherent limits, however [5]. Once a product is on the market, only small modifications to the manufacturing process and to the product itself are possible. Environmental effects can be reduced by reuse, recovery/recycling, or by incineration with heat recovery, but the most desirable alternative, material reduction, is no longer available at this stage.

5.2.1   INTEGRATED PRODUCT DEVELOPMENT

One of the earliest innovative approaches to product development with direct implication for design for the environment was product-aimed rationalisation based on the value analysis philosophy [5]. This systematic method for achieving the necessary functions of a product at the lowest possible cost starts, in contrast to traditional cost-reduction methods, by asking such fundamental questions as:

- What product is it?
- What is its function?
- What is the value of the main function?
- What alternatives are available to achieve this function?
- What is the cost of these alternatives?

Value analysis does not try to break down already established costs; instead, it seeks to build a new solution from a value established for the main function of a product, a reasoning similar to the determination of the functional unit in life cycle assessment. It follows that value analysis can elucidate hidden potential for environmental improvement that can directly translate into cost savings. As with LCA, it is likely that value analysis will be particularly effective in the development of new products rather than in the rationalisation of existing products.

To meet new requirements for product development, several approaches such as *concurrent engineering* [6, 7] and *integrated product development* [8] have been proposed. They emphasise the importance of early involvement of those responsible for the different stages of product design (Figure 5.4). This interdisciplinary structure has been widely adopted as a strategy for agile manufacturing, allowing companies to release higher-quality products while reducing time to market by addressing relevant product issues to as large an extent as possible during the concept and advance planning stage.

The relevance of these approaches to environmental issues is that they recognise the importance of considering the entire life cycle of a product and involving all those implicated in product development from an early stage. Product development procedures with their roots in integrated product development such as life cycle design (LCD), design for the environment (DFE), and Life Cycle Engineering (LCE) enlarge traditional decision making in engineering to include environmental aspects.

Design strategies such as Design for Manufacturing and Assembly (DFM&A), Design for Recycling (DFR), and Design for Disassembly (DFDA) give valuable information on how to improve materials separation for recycling, material use and product maintenance. Material selection methods are appearing that relate traditional mechanical and physical properties to environmental performance [9]. Common to all environmental design strategies is the preventive avoidance of problems. They can be used for the identification of weak points and of possibilities for improvement in existing systems [10].

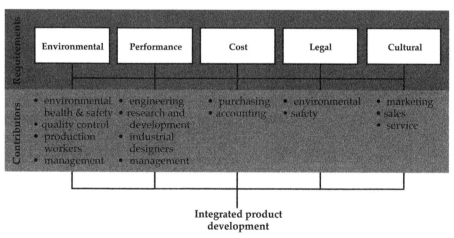

Figure 5.4   Schematic of integrated "green" product development.

5.2.2   COMPUTER-BASED "GREEN" DESIGN TOOLS

Since design for environment is a relatively new discipline, few designers have formal training in the field. Educational material, design guidelines and tools and verification methods of design decisions are needed. In product development, stringent time and economic constraints rarely allow designers time to study guidelines and regulations. It would be desirable to integrate environmental product regulations and data in computerised user-friendly design analysis tools and design support. A valuable source of information on these tools is the yearly International Seminar on Life Cycle Engineering, arranged by The International Institution for Production Engineering Research (CIRP) since 1993, where new computer-based design tools are presented each year. Table 5.1 presents a selection of currently-available software tools.

Table 5.1   Software-based design tools.

Name	Source	Application
ReStar	Carnegie Mellon University, USA [11]	Identification of optimal recovery path. Assessment of design for disassembly, recycling and repair.
Environmental Standards Processor	Carnegie Mellon University, USA	Search, selection, access and use of applicable standard provisions to evaluate product conformity.
Component Design Advisor	Carnegie Mellon University, USA	Assessment of potential product improvements based on design guidelines and rules of thumb.
Design for Environment Tool	Boothroyd Dewhurst Inc., USA and TNO, The Netherlands	Product analysis with respect to disassembly. Cost-benefit analysis for design optimisation
EcoDesign Tool	Manchester Metropolitan University/Nortel, UK [12]	Decision support for designers based upon expert rules.
ECO-Fusion	NEC Corporation, Japan [13]	Environmental product assessment in relation to a reference product, life cycle assessment, assembly/disassembly evaluation.
RecyKon	University of Erlangen, Germany [14, 15]	Design for recyclability based on material and energy flow charts for evaluation of product-solutions with assistance from external and internal data bases.

5.3 REDUCTION OF MATERIAL INTENSITY

Reducing material consumption is nothing new. While it has now come to make environmental sense, it has always made economic sense: less material means reduced material, storage and processing costs. It is of high priority for reducing environmental effects since it can be affective at the level of the product, the process and of waste.

5.3.1 MATERIAL SELECTION

New plastics and polymer-based composites are continuously being developed, and it is difficult for a designer to be aware of all material types, properties and selection criteria. In addition to cost, availability, performance, functional and manufacturing requirements, environmental aspects and recyclability are gaining importance in the selection of materials. The natural tendency to choose a known material rather than to look for new alternatives may lead a designer to propose sub-optimal solutions. For this reason, methods are being developed for systematic material selection for specific applications.

Ashby [16] has created material selection charts displaying property clusters for different material types by plotting one property against another (for example, modulus versus relative cost). These charts are overlaid with the contours for applications-based property indices (such as modulus over density times cost), as schematically displayed in Figure 5.5. This permits selection of materials according to application specifications. These charts have been developed for several important mechanical and thermal properties and for the easiest environmental effect to measure, namely energy. It would be feasible to expand these charts to include further environmental parameters, if quantitative measures of environmental impact were to be developed.

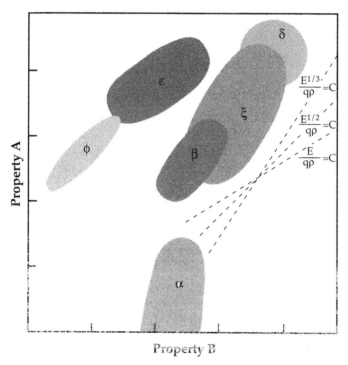

Figure 5.5   Schematic representation of an Ashby-like materials property chart.

Access to comprehensive materials data bases such as CAMPUS [17] and PLASCAMS [18] is also necessary if material selection is to be effective. Such data bases are often administered by one or several polymer producers. The CAMPUS-system [17], is jointly administered by BASF, BAYER, HOECHST and HÜLS. The development of such data bases to include environmental and traditional material data would significantly facilitate product development. One example of such a system is the ImSelection software developed at Carnegie Mellon University, USA, which integrates mechanical design criteria with life cycle environmental consideration in material selection[9].

5.3.2 WEIGHT REDUCTION

When weight reduction is a priority, the specific properties of materials (their properties divided by their density) can be of greater significance than their absolute properties. In Figure 5.6, the absolute and specific properties of a selection of materials is shown. Despite their relatively low intrinsic stiffness, plastics and polymer-based composites offer considerable potential for weight reduction, since their specific stiffness is high.

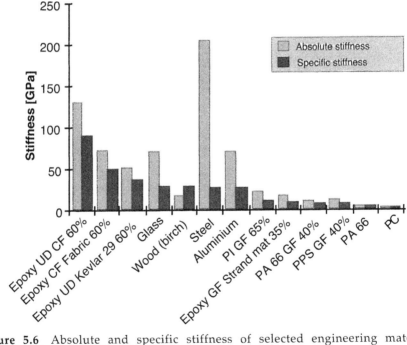

Figure 5.6 Absolute and specific stiffness of selected engineering materials (UD = uni-directional, GF = glass fibre, CF = carbon fibre).

The reduction in the weight of packaging has been made possible by the development of new materials fulfilling traditional requirements using less material. For instance, flexible packaging is now taking over applications that traditionally have been dominated by rigid packaging. Refill containers for milk and paint are two examples. The great advantage with flexible packaging is that it is easily compacted and easier to handle during transport. Semi-rigid packaging materials such as foamed PET glycol (PETG) or foamed PP are replacing solid materials in many applications. They offer better thermal insulation and lower specific weight [19]. The development of new plastics packaging materials for food wrapping has reduced the weight of the material required to wrap five kilos of beef from 62 grams in the 1970s to 29 grams in 1993 [20]. Similar developments can be seen in plastic soda bottles whose weight has been reduced by 25 % between 1977 and 1990 (Figure 5.7).

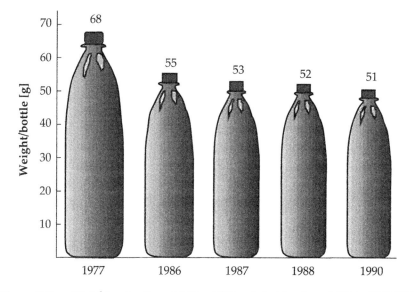

Figure 5.7   Weight reduction of 2-litre PET soda bottles from 1977 to 1990 [21].

Aside from weight reduction, the following the following alternatives can be used to reduce environmental impact and resource consumption per delivered service unit:

* concentration,
* refilling,
* recycling.

Concentration is achieved by changing the formulation of the product to deliver higher performance per unit volume. For example, instead of delivering one litre of ready-to-drink juice, a concentrate occupying a fifth of the volume can be delivered to the customer, who just adds water to it. The transported weight is reduced, but the same function is delivered to the customer.

Refilling is carried out on two levels, either by the producer, as in the refill of beverage bottles, or by the customer (for example flexible refill packages for rigid shampoo bottles). The environmental gain in using refillable beverage bottles is less obvious than in the case of shampoo bottles, since the refillable beverage bottles must be made thicker to resist several cycles of washing and transport.

In transportation applications weight reduction can have knock-on effects (Figure 5.8). Reducing the weight of components far from the centre of gravity is of much higher value since the corresponding improvement in handling and active security is much higher than if the weight were to be gained close to the centre of gravity.

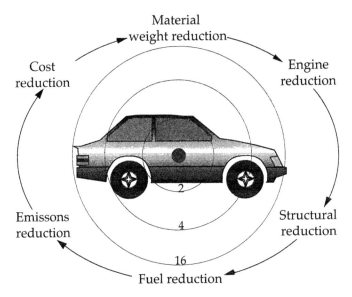

Figure 5.8 Value and synergy effects of weight reduction in transportation applications (after Månson [22]).

Such gains can lead to weight reductions in surrounding structural parts. In the case of a car, the engine could be redimensioned since it would need to power a lighter car, thus reducing fuel consumption and also the structure of the car (which would not need to support such a heavy load of component parts).

For these reasons plastics are increasingly being applied within the transportation industry. Examples of spectacular weight savings achieved through the replacement of metals with polymers in automotive applications are such as HDPE fuel tanks and battery boxes representing weight savings of up to 40% and 70% respectively [23].

Material reduction in processing

Reducing the material used in the product is not the only way of reducing material use. In some cases a considerable amount of material is lost in the form of process waste (for example, in-plant trimmings) or in processing aids such as solvents for painting. An example of how new technology can reduce material use is simultaneous injection moulding and painting within the mould, called GIPT (Granulated Injection Paint Technology) developed by the Rover Advanced Technology Centre at the University of Warwick, Evode Powder Coatings Ltd. and Battenfeld AG [24]. The technique is based on a specially-developed colour granulate that is injected initially to form the exterior skin. Immediately after this the core material is injected and integrated with the skin. The skin does not form flow lines on the surface, even if the core would do so, and surface quality can consequently be even better than for conventionally-painted surfaces. This method is of great interest for the automotive industry since it is capable of generating class A surfaces without post-treatment. Since it eliminates the painting step, costs are reduced significantly. Furthermore, all emissions linked to the use of solvents are eliminated. The complete encapsulation of the core in the skin allows the use of any colour for the core; thus recycled material, whether in-plant or post-consumer can be mixed in without affecting surface appearance. The first commercial application of this technique was in hubcaps for Rover [24].

In injection moulding processors often use multi-cavity moulds to make several products in one cycle. The runners required to supply material to each cavity contain a considerable amount of material which, if the product itself is small, may be of the same mass as the parts themselves. This is evidently quite uneconomic. Hot runner systems solve this problem by keeping the material in the runners from solidifying between moulding cycles.

Hollow profiles and foams

Gas-assist injection moulding is becoming a frequently-used process for forming hollow, stiff parts. Foamed material, sandwich structures with compatible core and face sheets, and sandwich constructions with recycled material are other examples of how intelligent material utilisation can lead to source reduction.

5.3.3 MATERIAL REDUCTION THROUGH DESIGN

Optimised wall thickness

Different design alternatives can be evaluated using computer-aided design (CAD) tools to minimise the wall thickness of a part without decreasing stiffness

[25]. An intrinsically stiffer material, a stiffer design of structure (using reinforcing ribs and bosses) or a combination of the two are possible alternatives.

Simplified design

Simpler design often involves less material. Simplification can be achieved through the omission of purely decorative details or through clever component design. Many of the obstacles for efficient recycling are overcome by the consolidation of parts into fewer sub-assemblies. Quick disassembly and high material recovery are crucial for economically viable recycling, and they are both aided by integrating functions into fewer parts. A good example is the use the computer industry has made of the design freedom of thermoplastics to reduce the number of parts and materials and the weight of computer housing [26].

Miniaturisation

The development of electronics equipment is a good example of how material consumption can be reduced by miniaturisation. Size reduction is mainly due to new developments within the field of electronics, allowing designers to make their products smaller, thus consuming less material per functional unit. However, making products smaller may pose problems for recycling and repair, since disassembly may become more difficult and time consuming. Well-performed miniaturisation must allow for rapid disassembly to avoid unnecessary added cost to the life cycle of the assembly.

Integrated functions

Resource use can be reduced by developing products which perform several functions simultaneously [27]. Examples of this are TV sets with integrated VCR or the integrated telephone, answering machine, and fax machine, the size of which now approaches that of a larger telephone.

Material and process integration

Thermoplastics offer a wide variety of properties ranging from the low intrinsic stiffness and high design freedom of neat and short fibre reinforced thermoplastics to the high intrinsic stiffness and low design freedom of continuous fibre reinforced thermoplastic composites. Different polymeric materials are often combined or are used in conjugation with metals to meet multiple application requirements. Traditionally, such material combinations have been achieved without taking into account end-of-life issues such as recyclability.

Most environmental design guidelines advise against combining materials in an irreversible way unless they are compatible with each other. Thus, material combinations such as those found in traditional automotive dash boards (a metal frame clad with foam with a softened PVC surface) would not be desirable.

The versatility of plastics, however, makes it possible to achieve similar performance by combining variants of only one base plastic: modification of the polymer chemistry of stiffening with a range of reinforcement configurations provides a range of properties. By carefully placing highly tailorable intrinsically stiff material, load transfer can be optimised, while the shaping freedom of neat or short-fibre reinforced thermoplastics can be used to provide geometry, surface performance, and functional integration (inserts, snapfits, etc.). This is shown in Figure 5.9. In this way the recyclability of the assembly is guaranteed while respecting the traditional performance requirements of the part.

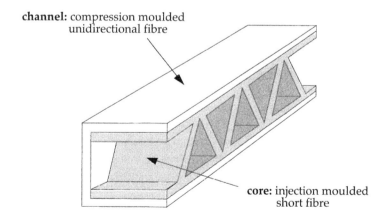

Figure 5.9   Beam with compression moulded unidirectional fibre channel section and short fibre reinforced injection moulded core [28].

In addition to respecting traditional performance criteria and assuring recyclability, the use of the same matrix materials in all parts of a sub-assembly offers the possibility of optimising the process. Traditionally, the manufacture of a two-part assembly such as that shown in Figure 5.9 requires three main operations: the processing of each part, and their subsequent bonding together. The transfer of parts between machines and finishing operations further complicate the process.

By using similar materials, the process could be integrated into one sequence [28]. The channel would first be pressformed and the short-fibre reinforced core would then be injected. Between these two operations only the male half of the mould would need to be changed, since the channel would serve as the new female "mould" for the injection. There would be no significant cooling of the channel between the operations. This not only reduces the energy consumption associated with each processing cycle and the overall cycle time (Figure 5.10), but also eliminates the need for surface preparation and storage between processing steps as well as ageing of the polymer during processing and intermediate storage. Furthermore, it reduces the thermal loading of the material during processing, thus improving the longevity of the material in service.

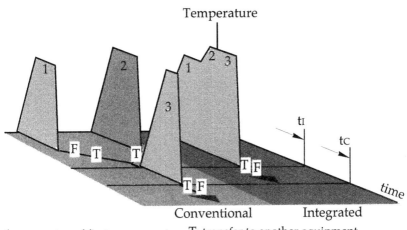

1: processing of first component   T: transfer to another equipment
2: processing of second component F: finishing operations
3: bonding t_I: cycle time for integrated processing
                                    t_C: cycle time for conventional processing

Figure 5.10 Comparison of processing cycles for conventional and integrated processing of a two-component part [29].

A reduction in material intensity and in waste generation could have a negative effect on recycling companies, who are dependent on a sufficient supply of material to operate economically. Unless the recovery infrastructure and the design of products are changed to allow a higher percentage of material recovery per product their business would suffer.

5.4 PRODUCT LIFE EXTENSION

The expression "plastic age" which is frequently used to describe the present era, has a connotation of continuous and wasteful acquisition of new objects. There is a popular belief that manufacturers are consciously building obsolescence into their products, and deliberately promoting customer dissatisfaction with existing products by introducing needless innovations [27, 30]. The issue of life extension is receiving increased attention. But can the demand for new products and the quantity of discarded products be decreased by extending the lifespan of a product? The "built-in" durability provided by the manufacturer is not a guarantee for a certain calculated life; products can be retired from service for several other reasons than normal wear and tear, such as [31, 32]:

- technological obsolescence,
- style obsolescence,
- damage caused by inappropriate use or accidents.

In many applications the intrinsic life of the material may be considerably longer than the life of the product itself., as can be seen in Figure 5.11. For example, who does not throw away perfectly functioning plastic grocery bags instead of reusing them at the supermarket, or plastic cups after a cup of coffee? In other applications such as leisure and industrial goods the product lifespan is more likely to approach the life of the constituent materials.

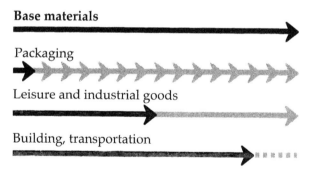

Figure 5.11   Durability of base materials versus durability of products.

In products undergoing rapid technological development, such as communication equipment and computers, it makes little sense to extend the service life beyond the period of technology renewal, and the producers may have little influence over the obsolescence of the product. The automotive industry is reluctant to extend vehicle life to more than 15-20 years, due to fashion as well as technological obsolescence.

Despite the fact that consumers are calling for more durable products, there are several cases where the increased purchase price of more durable goods deter the consumer, even though the total cost during the useful life of the product may be lower. Before embarking upon product life extension it is therefore important to be aware of and understand the consumer patterns for the product.

If product life extension is to become an interesting alternative for producers it is important that they maintain better control over their products. This can be achieved by, for example, product leasing. By selling the service that the product offers, producers will obviously be more willing to design their products for maintainability, serviceability and longer life. One possible scenario would be that producers of communication equipment make a profit on the intensity of use of their product, such as charging a fee per transferred page in the case of fax producers [33].

5.4.1 MODULARITY AND MAINTAINABILITY

In complex assemblies, sub-assemblies or individual parts may degrade at different rates. If such a product is subject to rapid technological change then it may be more appropriate to design in a modular form which permits replacement and upgrading of obsolete parts in a way that is *adaptable* to the main structure. Computers with upgradable hardware are designed in this manner. Firm control of tolerances and dimensions of the manufactured parts is a prerequisite for this approach [32].

A product designed for a long life normally requires maintenance to retain optimum performance. To this life potential it is necessary to design products in a way that makes maintenance simple, fast, and cost-effective. The balance between the price of a new product and maintenance costs can be decisive for when it comes to retiring a product. The maintainability of a product can be measured by the time required to restore it to a specific condition in accordance with prescribed procedures and resources [34].

When designing a product for maintainability, several factors should be taken into account:

- Who will maintain the product?
- What is their level of expertise?
- What tools are available for maintenance?
- What are the type and quantity of parts required for the maintenance units to operate efficiently?

Good maintenance requires design for assembly and disassembly. Short disassembly and reassembly times are crucial since the economy of maintenance is a function of the maintenance down time. Design for maintainability requires an evaluation of the parts requiring most frequent maintenance and adaptable design to allow easy access for inspection, repair, and replacement. Typically, replacement of screw fasteners with quickly disassembled fastening devices such as two-way snapfits (Figure 5.12) reduces disassembly time considerably. The disassembly points can be indicated by moulded-in logos which can also indicate the type of tool to be used and the appropriate disassembly procedure.

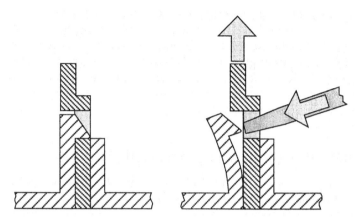

Figure 5.12   Two-way snap fit, easily disassembled with a screwdriver.

Labour costs often constitute the lion's share of maintenance costs. Service will be significantly simplified if the product is simple, and wear parts are made easily accessible for the maintenance personnel and adapted to the tools they have available.

5.4.2   REPAIR, REMANUFACTURING AND REUSE

When a durable product is retired from service it may be possible to restore it, or at least certain parts, to an "as-new" condition through remanufacturing. After full disassembly of the product, usable parts are cleaned, refurbished and put into inventory. Then, a new product can be assembled using both new and old parts. The performance of this product can be similar to that of a new product. Examples of parts frequently remanufactured are car starter assemblies and manufacturing equipment such as machining tools [35].

Remanufacturing offers significant potential for material and energy savings as well as the reduction of waste, since

- the consumption of natural resources is reduced;
- the energy used to extract and refine the raw material is saved;

- the energy consumed in processing and finishing the product is saved;
- the energy consumption of recycling operations is saved;
- material is diverted from recycling and disposal.

It must be kept in mind that the remanufacturing operations are only justified when a used product can be repaired to as high a standard as the original equipment. Also, if the overall cost balance is negative, there is obviously no reason to remanufacture. Furthermore, a functioning remanufacturing scheme depends on the availability of discarded products, a low-cost trade-in network, inventory infrastructure, and a constant demand for the remanufactured product. If a product is likely to be unattractive to customers for reasons of style or due to technical obsolescence, then it is not suitable for remanufacturing. Customer prejudice towards remakes is often an obstacle to remanufacturing especially within the consumer product industry. This is likely to change; owning a rebuilt product may become a sign of environmental consciousness in the near future.

There are already cases where remanufacturing has become a profitable business, even within an industry subject to such rapid change as the computer industry. In many cases public institutions not having the resources to replace their entire stock of computers sign contracts with computer manufacturers to upgrade their machines. In such cases, the product manufacturer has obvious advantages to gain by designing the product for maintainability to allow safe and fast dismantling and replacement of parts.

Products can in some cases be returned to service without remanufacturing. Milk bottles are returned, washed and refilled numerous times before being ultimately retired from service. Another example is reusable transport packaging in the form of pallets and boxes; the use of plastics in pallets increases the number of times a pallet can be used before being discarded in comparison to wood pallets. Transport packaging producers are forming packaging pools which allow them to have greater control of their packaging. Often the pools use a deposit system to maintain the packaging within the pool (Figure 5.13).

Reuse becomes an interesting alternative for products where the stages before and after use have a dominating effect on the environment. The waste intensity per service for such a product is directly proportional to the number of cycles. On the other hand, in cases where the service phase is found to have the major impact according to a life cycle assessment, reuse will have little or no effect on the overall impact.

A key factor for successful reuse is, just as in recycling, an effective reclamation system. This reclamation system can be used to recover retired products into a recycling scheme.

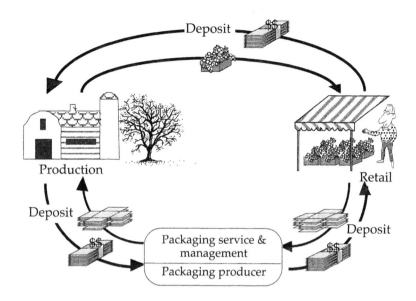

Figure 5.13   Transport packaging and financial cycle of a pool system with deposits.

A product designated for reuse has to be designed to resist refurbishing operations such as cleaning or replacement of worn parts. This may signify increasing wall thickness relative to the single-use product or choosing a more resistant material. The changes made in adapting a product for reuse must be taken into account in the environmental balance. As discussed in Chapter 2, thorough knowledge of the properties of plastics will allow improved material utilisation and improved control of the durability and reliability of products through appropriate design.

5.5 MATERIAL LIFE EXTENSION

While prolonging product life is an indirect way to extend material life, material substitution, reformulation and recycling can also be useful. The performance value of the material can be kept at a high level during service by careful matching of the material properties and application requirements. This ensures recycling with as high recycled properties as possible.

5.5.1 MATERIAL OPTIMISATION

As a rule of thumb, given that a material fulfils all relevant functional demands, the material that has the lowest environmental impact should be chosen. This generally means that the product, or at least the material, has to be recyclable, since this diverts materials from the waste stream. The evaluation of the environmental impact is based on such factors as lifetime, performance, weight and the life cycle. Within the automotive industry, for example, life cycle assessment has identified fuel consumption during use as the dominant factor influencing the environmental impact; thus lightweight design is accorded higher priority than recyclability.

Often the desired functions cannot be provided with one single material, and both material selection and assembly become crucial for successful recycling of the product. If two materials are to be assembled by means of an irreversible bond, such as welding or an adhesive, it is important to select both compatible materials and a compatible adhesive to avoid an unnecessary drop in material quality upon recycling (see Figure 5.14 : "poor assembly").

If compatible materials and adhesives are selected to give good recycled properties without the need for separation, then the cost of recycling is reduced. When compatible materials cannot be found that meet requirements for a given function, the assembly method can be adapted to provide easy material separation with; for example, reversible snapfits or water-soluble adhesives. The biopolymer industry is actively searching for water-soluble glues for the bonding of incompatible materials in multilayer beverage cartons, which could significantly facilitate material separation. This is predicted to become an enormous market for those manufacturers who first bring a suitable product to market [36].

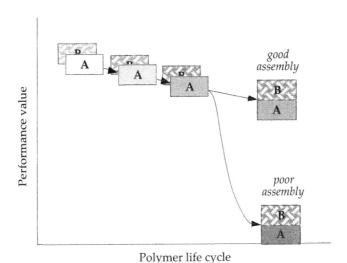

Figure 5.14 Reduction of material performance value due to improper material selection or choice of assembly method.

5.5.2 DESIGN FOR RECYCLING

The majority of products recovered nowadays have not been designed with the recovery of their material constituents in mind. Recycling is a valuable way to divert material from the waste stream. It is well known, however, that recycling operations also cause environmental impacts, in some cases greater than the environmental and economic gain of the operation. It is important for designers to be aware of the requirements of a functioning recycling network before embarking upon designing products for recycling.

The main considerations in Design for Recycling (DFR) are:

* fewer materials,

* fewer sub-assemblies,

* design for easy and efficient disassembly,

* material identification systems,

* specification of recyclable materials for products,

* specification of recycled materials for products.

Reducing the number of materials in an assembly can significantly improve the possibilities of prolonging the life of the material. The recycled properties of single material structures are easier to assure than those of commingled blends. The joining of two or more materials forms an interfacial layer that normally has to be considered as a third material with specific properties. Furthermore, the state of the constituent materials and their stress state change, making it difficult to predict the service performance and recycled performance of the material combination (Figure 5.15). The performance during the first service life may be improved, but if incompatible materials are combined in an irreversible way it becomes difficult to attain sufficiently good properties for a useful recycled application.

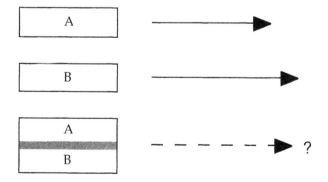

Figure 5.15 Material durability versus product durability for plastic and composite assemblies.

A comprehensive guide containing information on design for recycling has been developed by the German association "Verein Deutscher Ingenieure" (VDI) [37].

Using fewer sub-assemblies often decreases the disassembly time per unit weight of recovered material and increases material capture. It also reduces the complexity of the product and leads to improved reliability since the reliability of a product depends on the number of components, their configuration and their individual reliability [32].

Currently-used disassembly methods are not necessarily intended for the recovery of plastics, since their recycling is still fairly new and economically unfavourable. The disassembly profile for a compact car, displayed in Figure 5.16 [38], nevertheless shows that within 20 minutes, 62% of all plastics, or 71% of the mechanically recyclable plastics, can be recovered. A total of 49 minutes was required to disassemble all plastic items and complete functional units. A further 41 minutes were required for separation and sorting according to plastic type. The plastics recovered during the first 20 minutes (30 kg) cost 2.50 DEM/kg to reprocess, while virgin material costs roughly 2.30 DEM/kg. On a strictly economical scale the recycled material will not be competitive, unless the disassembly, separation and sorting costs are reduced.

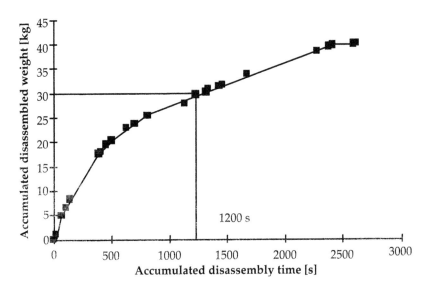

Figure 5.16   Disassembly data for a compact car (source FAT Schriftenr. 100 [38]).

Another source of design aids is the "Environmental Considerations in Product Design and Processing" [26], which contains guidelines for design for disassembly and recyclability, design for reusability, and waste avoidance during processing. It gives priority to design for recyclability and disassembly over design for reuse. The latter may be of higher priority if the production and disposal of a product creates a significant share of the environmental impacts. The guidelines stress design innovation, common sense, and material knowledge as key factors in achieving good design for recyclability and disassembly.

> **Design for Recycling:**
>
> **Design innovation + Common sense + Material knowledge**

A hierarchy of strategies is established according to the following:

1   Source reduction by part consolidation and weight reduction.

2   Use of recyclable materials, such as thermoplastics instead of thermosets.

Furthermore the following design issues are discussed:

- use of compatible materials, adhesives, coatings, and labels on products (if this is not possible, make incompatible materials easily separable);

- avoid the use of paper labels since paper has a detrimental effect on the recycled properties of plastic materials;

- avoid metal inserts, or make them easily separable by the use of predetermined break points or break-out inserts;

- avoid unnecessary screws, metal clamps and over-moulding of plastics onto other materials;

- use snap-fits to enable quick separation and avoid other materials that may deteriorate properties; two-way snap-fits replace screws in parts that are disassembled frequently and snap locks can be used for parts that are permanently joined together;

- provide easy identification by bar-coding materials or by moulded-in logos.

These comprehensive guidelines give an idea of how the quantity of material for recycling and the quality of recycled material can be increased through proper design of products.

It must be kept in mind that design for manufacture and assembly and design for disassembly may sometimes be in conflict with each other. For example, snap-fits may accelerate the assembly process but impede rapid disassembly if they are not designed properly. Furthermore, safety rules and regulations, such as for consumer electronics and household appliances, require the protection of electrical circuits to avoid accidents. In other products, easy access may also encourage the theft of valuable components [32].

5.6 PERSPECTIVES

As design for resource optimisation becomes important, so does the link between product development and material/process optimisation. While practical and intuitive guidelines exist, there is still some way to go before systematic computer-based tools reach widespread acceptance within product development groups. Most current design software concentrates on design for manufacturing, assembly and disassembly and material selection. What is lacking is input about material life cycle performance issues such as reliability and durability. Such knowledge is necessary in design support software in order to assure successful development of more environmentally-benign products.

The implementation of integrated product development allows environmental and cost issues to be approached in a more systematic and innovative way, and has been proven to give product performance and cost advantages. Much development within the field of Life Cycle Engineering is still to come. It is likely that technology will be developed that will help solve problems of waste by minimising resource use. The increased emphasis on environmental issues expressed by consumer organisations and governments not only shifts the focus onto product performance but also onto the environmental performance of entire organisations. The issue of environmental management will be covered in the following chapter.

REFERENCES

1.   R. Züst, *Sustainable Products and Processes* in proceedings of *ECO-Performance '96*, Zürich, Switzerland, Verlag Industrielle Organisation, Zürich, pp. 5-10 (1996).

2.   J. Fiksel, *Design for Environment: an Integrated Systems Approach* in proceedings of *IEEE International Symposium on Electronics and the Environment*, Arlington, VA, USA, IEEE, Piscataway, NJ, USA, pp. 126-131 (1993).

3.   W. J. Glantschnig, *Green Design: An Introduction to Issues and Challenges*, IEEE Trans. Compon., Packag. and Technol. Part A, **17**, 4, pp. 508-513 (1994).

4.   B. S. Blanchard, *System Engineering Management* , John Wiley & Sons, Inc.: New York, p. 8 (1991).

5.   L. D. Miles, *Value Engineering: Wertanalyse, die Praktische Methode zur Kostensenkung*, 3rd ed., Moderne Industrie: Munich (1969).

6.   J. R. Hartley, *Concurrent engineering: Shortening Lead Times, Raising Quality and Lowering Costs*, Productivity Press: Cambridge, MA (1992).

7.   A. Kusiak, *Concurrent Engineering: Automation, Tools and Techniques*, John Wiley & Sons, Inc.: New York (1992).

8.   M. M. Andreasen and L. Hein, *Integrated Product Development*, IFS Publications Ltd.: Berlin (1987).

9.   R. W. Chen, D. Navin-Chandra, I. Nair, and F. Prinz, *ImSelection-An Approach to Material Selection that Integrates Mechanical Design and Life Cycle Environmental Considerations* in proceedings of *IEEE International Symposium on Electronics and the Environment*, Orlando, USA, IEEE, Piscataway, NJ, USA (1995).

10.   K. Saur, J. Gediga, J. Hesselbach, M. Schuckert, and P. Eyerer, *Life Cycle Assessment as an Engineering Tool in the Automotive Industry*, The International Journal of Life Cycle Assessment, **1**, 1, pp. 15-21 (1996).

11.   D. Navinchandra, *ReStar: A Design Tool for Environmental Recovery Analysis* in proceedings of *9th International Conference on Engineering Design (ICED '93)*, Den Haag, The Netherlands, pp. 780-787 (1993).

12.   J. R. Poyner and M. Simon, *The Continuing Integration of the Ecodesign Tool With Product Development* in proceedings of *IEEE International Symposium on Electronics and the Environment*, Dallas, USA, IEEE, Piscataway, NJ, USA, pp. 201-206 (1996).

13.   S. Miyamoto, T. Tamura, and J. Fujimoto, *ECO-Fusion Integrated Software for Environmentally-Concious Production* in proceedings of *IEEE International Symposium on Electronics and the Environment*, Dallas, USA, IEEE, Piscataway, NJ, USA, pp. 179-184 (1996).

14.   H. Meerkamm and J. Weber, *Integration of the Design for Recyclability-Tool RecyKon in an Environmental Management Concept* in proceedings of *ECO-Performance '96*, Zürich, Switzerland, Verlag Industrielle Organisation, pp. 159-166 (1996).

15.   K. Feldmann, H. Scheller, H. Meerkamm, and D. Krause, *Design for Recyclability and Economic Planning of Disassembly Based on the Recyclinggraph Tool* in proceedings of *RECY '94*, Erlangen, Germany, pp. 76-90 (1994).

16.   M. F. Ashby, *Material Selection in Mechanical Design*, Pergamon Press Ltd.: Oxford, UK (1992).

17.   CAMPUS, *Computer Aided Material Pre-Selection By Uniform Standards*, BASF, BAYER, HOECHST, HULS AG (1990).

18.   PLASCAMS, Rapra Technology Ltd.

19.   J. Myers, *Drive for Lightweighting in Plastics Packaging Intensifies*, Modern Plastics International, **24**, 9 (1994).

20.   V. Williams, *Plastic Packaging for Food: The Ideal Solution for Consumer Industry and the Environment* in proceedings of *R'95*, Geneva, Switzerland, **2**, pp. 49-56 (1995).

21.   R. F. Testin and P. J. Vergano, *Packaging in America in the 1990s*, Clemson, SC, USA (1990).

22.   J.-A. E. Månson, *New Demands on Manufacturing of Composite Materials* in proceedings of *High Performance Composites: Commonalty of Phenomena*, Rosemont, Illinois, USA, The Minerals, Metals & Materials Society, pp. 3-19 (1994).

23.   J. Maxwell, *Plastics in the Automotive Industry*, Woodhead Publishing Ltd.: Cambridge, UK, p. 7 (1994).

24.   *"Formsprutlackering" ger Färdigmålad Detalj*, Plastforum, 12, p. 9 (1993).

25.   J. Murphy, *Konstruktion för Återvinning: Morgondagens Lönsamhet Bestäms Idag*, Plastforum Scandinavia, 1 / 2 (1996).

26.   *Environmental Considerations in Product Design and Processing*, GE Plastics.

27.   D. McKenzie, *Design for the Environment*, Rizzoli International Publications, Inc.: New York, p. 77 (1991).

28.   G. D. Smith, S. Toll, and J.-A. E. Månson, *Integrated Processing of Multi-Functional Composite Structures* in proceedings of *39th International SAMPE Symposium*, Anaheim, **2**, p. 2385 (1994).

29.   P. E. Bourban, F. Bonjour, and J.-A. E. Månson, *Automated Integration of Materials and Processing techniques for multi-functional composite parts* in proceedings of *ECCM 7*, London, UK, Woodhead Publishing Ltd., **1**, pp. 201-206 (1996).

30.   J. Butlin, *Product Durability and Product Life Extension: The Contribution to Solid Waste Management*, Organisation for Economic Cooperation and Development (OECD), OECD Publication Office, Paris, France (1982).

31.   M. A. Moss, *Designing for Minimal Maintainance Expense: The Practical Application of Reliability and Maintainability*, Marcel Dekker Inc.: New York, p. 48 (1985).

32.   G. A. Keoleian and D. Menerey, *Sustainable Development by Design: Review of Life Cycle Design and Related Approaches*, Air & Waste, **44**, May, pp. 644-668 (1994).

33.   O. Giarini and W. R. Stahel, *The Limits to Certainty: Facing Risks in the New Service Economy*, Kluwer Academic Publisher: Amsterdam, The Netherlands, p. 83 (1993).

34.   US MIL-Std-721 C.

35. R. T. Lund, *Remanufacturing*, Technol. Rev., **87**, pp. 19-24 (1984).

36. P. Krusell, *Säckar och Bestick för Komposten: Starkare Nedbrytbar Plast ska ta Marknadsandelar*, Plastforum Scandinavia, 12, p. 43 (1993).

37. *Konstruieren Recyclinggerechter Produkte*, Ver. Deutsch. Ing., VDI 2243 (1991).

38. *Demontage und Verwertung von Kunststoffbauteilen aus Automobilen*, Forschungsvereinigung Automobiltechnik EV (FAT), Schriftenreihe Nr 100 (1993).

6

ORGANISATIONAL ASPECTS OF LIFE CYCLE ENGINEERING

The implementation of technical solutions to environment-related problems requires organisational change. Environmental Management Systems (EMS) allow companies to systematise their approach to environmental issues by identifying environmental liabilities and by improving efficiency as well as creating business opportunities. The ICC Business Charter for Sustainable Development, the British green audit Standard BS 7750, the European Community Environmental Management and Audit Scheme and the international standard series ISO 14000 are reviewed. Adhering to these standards is not free from cost, but long-term competitive advantages such as access to markets, finance and insurance as well as improved operational efficiency are likely to compensate for initial costs. Opportunities for support in the implementation of EMS for small and medium-sized companies are discussed.

6.1 THE IMPORTANCE OF NETWORKING

The Life Cycle Engineering concept as it applies to technical solutions to environment-related problems within the plastics industry has so far been discussed. It must not be forgotten that this technical development takes place in a much broader context. Companies exist and interact within a social, political, economic and corporate framework that provides business both opportunities and finance, but which also places restrictions on corporate activities through legislation, taxation, public pressure, competition, and market demand.

There are strong arguments for polymer producers to become environmentally aware. Increased public and legislative attention to environmental issues has obliged industry to review its role as not only an actor in the economic system, but also as a part of the ecological system. It is not always easy for a company to evaluate its activities in the light of environmental concerns. Technical solutions are essential to resolve shortcomings but should be seen as only one aspect of business organisation and logistics. How a company manages environmentally-induced change can be as important as the technical nature of the change itself.

As an example of the complexity of organisational issues associated with Life Cycle Engineering, Figure 6.1 shows the possible stakeholders in a recycling scheme of an automotive under-the-bonnet application, in this case a radiator end-cap made from a reinforced polymer, short glass fibre-reinforced polyamide 66 (PA 66 + GF).

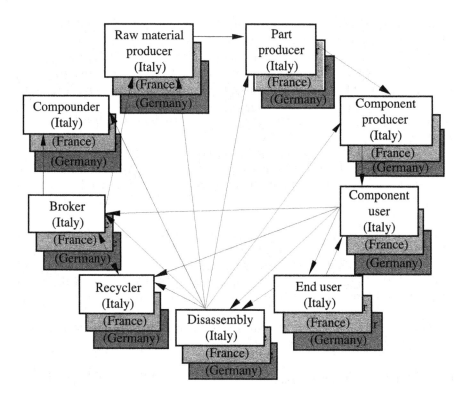

Figure 6.1   Relations and possible material flows within a an automotive recycling scheme.

When a radiator assembly reaches the end of its service life, a series of dismantling operations reduces the complex assemblies to their constituent parts. Each country will have a chain of operators involved in a national recycling network, and this example focuses on one country, Sweden. Once the used radiator end-caps are available, they can be sent for reprocessing to the company responsible for reprocessing. But of all the companies involved in the original production chain, from the raw material supplier to the component sub-supplier and finally the car producer, which is to be held responsible?

The arrows in Figure 6.1 show the multitude of possible routes the scrap material could take and the subsequent distribution for reprocessing, as the situation was perceived in the early 1990's. Logistics systems were ill-adapted to deal with the complex movements of used material back up the original chain of production. Reverse distribution costs could account for costs up to nine times higher than supplying the original product to the consumer, since returned goods often cannot be handled and stored in the same manner [1].

The actors in these material recovery chains tend nowadays to join forces to create viable reverse distribution structures. Constraints on the routes available certainly exist, either because of legislation or quality control requirements. For example, the quality of PA 66 + GF is related to fibre length, which is inevitably affected by compounding operations. Consequently, reverse routes going from part producers to compounders are possible from a technical point of view only if fiber length reduction can be controlled. Recovering material through the automotive producer is not particularly favoured either, since the recovery and recycling of products is not the core business of the original equipment manufacturer (OEM). It is, however, likely that the OEM would be held responsible for the recovery and recycling of its product. Draft take-back regulations are being introduced in several European countries, of which Sweden [2].

Such regulations define producer responsibility in development, manufacturing and marketing of products designed to be recovered or disposed of in an environmentally compatible way. Responsibility for using secondary raw materials in the manufacture of products and for ensuring material collection, material identification, reuse and recovery is also covered.

Figure 6.1 shows clearly that the issue of producer responsibility is much more complicated than draft legislation would lead to believe.

Responsibility also implies accountability. There are several aspects of the take-back operation over which producers may have little or no influence, namely:

- logistics,
- administration and cost of take-back programmes,
- disposal issues,
- reuse of dismantled products and use of secondary material,
- establishing of an effective end-of-life industry,
- environmental burdens of products already on the market.

The production processes may be local, but products are global. Cars are being produced at a limited number of locations, but they are used everywhere.

It is difficult to imagine any producer feeling comfortable with the idea of accountability without influence. There are, however, several other parties with a vested interest in the fulfilment of producer responsibility. If a producer is taken out of business because of non-compliance with regulations, then sub-suppliers, service and maintenance companies, refurbishing firms and end-of-life industries all lose a client. All these actors in the life cycle chain can be motivated to contribute to a solution to the end-of-life problem.

It is clear that the whole chain must contribute to the economy of operations by establishing targets for recovery and recycling rates, developing product design, dismantling and recycling technology, as well as take-back logistics. The main task is to establish who is responsible for which end-of-life operation. This is of course closely linked to the cost-benefit balance for each actor.

Examples of OEM-controlled recovery co-operations have appeared within several sectors such as the automotive industry, durable goods, information technology equipment, and carpets [3-5].

6.2 THE IMPORTANCE OF COMMUNICATION

There is increasing demand for information on environmental activities of companies. Governments, environmental pressure groups, and more importantly, investors, insurance companies and clients are beginning to include environmental performance as part of their evaluation of company quality.

In Figure 6.2, some of the parties interested in and affected by company environmental performance are shown. They all have different relations with the company and different ways of communicating and interpreting information.

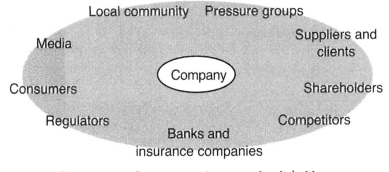

Figure 6.2   Company environmental stakeholders.

Effectively managing and communicating company environmental performance to such a wide variety of receptors requires some form of systematisation. This can take the form of Environmental Management Systems (EMS), Environmental Accounting (EA), Environmental Performance Indicators (EPI) and environmental reporting.

Managing environmental issues is not only done to satisfy external pressure for environmental change. There are several business opportunities in doing so, according to the authors of the standards and guidelines which will be reviewed in this chapter. Frequently-cited opportunities are:

- rationalisation and efficiency, reduction of waste,

- competitive position (products, processes, entering new markets before competitors),

- improved public image, attracting the brightest people,

- public acceptance,

- relationships with enforcement authorities (permits, controls, inspections, reporting),

- legal security (vis-à-vis mandatory requirements),

- civil liability,

- prevention of unforeseen costs and economic penalties,

- access to finance,

- access to insurance,

- shareholder confidence and added value of share price.

The last three points deserve particular attention: industrial stakeholders such as governments, public interest groups, environmental, legal and financial consultants, as well as banks and other financial institutions are increasingly asking for information on industrial environmental performance when making investment decisions and selecting products.

Ethical investments and so called "Green Funds" are likely to make compliance with EMS a necessity for attracting investment. Funds for environmentally-sound investments are rapidly emerging on the market [6-9]. So far they seem to attract mainly small or private investors, but in the long run larger investors will almost certainly pay more attention to this type of investments. Banks and financial institutions are also facing public scrutiny for environmental performance of their investments. Several banks are increasingly applying environmental audits on companies before granting loans and financial institutions are appearing which seek to take advantage of expanding markets in the environmental field by introducing environmental criteria into their investment strategy [10].

Long-term environmental performance improvement is increasingly being interpreted as an aid to long-term profitability and competitivity. Insurance companies are well aware of the risk of poor environmental performance of their clients. Companies with effective environmental management can demonstrate that they represent a lower risk and can use this in negotiation for lower premiums. Similarly, banks run risks when taking over businesses as collateral for unpaid loans, since they also take over the responsibility for operations and are thus liable for any kind of environmentally unsound practices in the company.

6.3 ENVIRONMENTAL MANAGEMENT

To manage environmental issues requires a company to establish clear structures and guidelines. Rather than proceeding in a hazardous manner, industry is systematising environmental management in a manner similar to quality management by considering features such as:

* environmental policies,
* environmental programmes,
* environmental management systems,
* environmental auditing procedures,
* environmental reporting procedures.

The company in question establishes and implements an environmental policy, site-specific environmental programmes and an environmental management system, carries out periodic site-specific environmental audits and prepares communication on environmental performance to certifiers, governments and the public.

Of the various guidelines, regulations and standards laid down for environmental management, the following sections review some of the most frequently used within Western Europe as well as the international standard ISO 14000.

6.3.1 THE ICC BUSINESS CHARTER FOR SUSTAINABLE DEVELOPMENT

The ICC Business Charter for Sustainable Development (1991) [11] is a proposal from the International Chamber of Commerce (ICC) to help businesses in improving environmental performance. The Charter was one of the first thorough publications on environmental management issues. It strongly reflects the growing importance of the concept of sustainability to businesses. The

---

Summary of the ICC Principles for Environmental Management

---

1   **Corporate priority:** recognising environmental management as among the highest corporate priorities, establishing policies, programmes and practices allowing environmentally sound conduct.

2   **Integrated management:** fully integrating policies, programmes and practices into each business as an essential element of management.

3   **Process of improvement:** continuous improvement of policies, programmes and practices taking into account technical developments, scientific understanding, consumer needs and community expectations, with legal regulations as a starting point.

4   **Employee education:** educate, train and motivate employees to act with environmental responsibility.

5   **Prior assessment:** assess environmental impacts before starting new activities and discontinuing or selling activities.

6   **Products and services:** developing products and services with as low environmental impact as possible and as high safety as possible whilst efficient in their use of energy and natural resources: recyclable, reusable and safely disposable.

7   **Customer advice:** to advise and educate customers, distributors and the public in the safe use, transportation, storage and disposal of products and services provided.

8   **Facilities and operations:** develop, design and conduct activities with respect to energy and material efficiency, sustainable use of renewable resources, minimisation of environmental impacts and waste generation and safe waste disposal.

9   **Research:** conduct or support research on the environmental impacts of raw materials, products, processes, emissions and waste associated with the enterprise to identify means of reduction.

10 **Precautionary approach:** modifying manufacture, marketing or use of products or services or the conduct of activities consistent with scientific and technical understanding, to prevent serious or irreversible environmental degradation.

11 **Contractors and suppliers:** promoting the adaptation of the principles with contractors acting on behalf of the enterprise, requiring improvements when appropriate as well as encouraging adoption by suppliers.

12 **Emergency preparedness:** develop and maintain emergency plans in conjunction with emergency services, relevant authorities and the local community in areas where significant hazards exist.

13 **Transfer of technology:** contribute to the transfer of environmentally-sound technology throughout the industrial and public sectors.

14 **Contributing to the common effort:** contributing to the development of public policy and to business, governmental and intergovernmental programmes and educational initiatives aiming to improve environmental awareness and protection.

15 **Openness to concerns:** foster openness and dialogue with employees and the public.

16 **Compliance and reporting:** measure environmental performance, conduct regular environmental audits and assessments of compliance with company requirements, legal requirements and with the ICC principles as well as providing this information to the board of directors, shareholders, employees, authorities and the public.

sixteen principles for environmental management (which, for business, is a vitally important aspect of sustainable development) constitute the core message of the ICC Charter.

The ICC's objective is for the widest range possible of companies to commit themselves to improving their environmental performance in accordance with these principles, to implement management practices to this end, to measure their improvements, and to communicate them to all relevant stakeholders.

The Business Charter encourages firms to be pro-active on environmental issues, anticipating legislation and setting the examples by carrying out research on environmental issues of concern in addition to the necessary legal compliance. The importance of communicating environmental activities to improve business, performance, and reputation is emphasised.

ICC cooperates closely with the UN Environment Programme (UNEP) on the subjects of environmental reporting, environmental management systems and training, technology assessment, and in providing advice for the implementation of the ICC Charter.

To meet the challenge of economic growth within the limits of sustainability, the ICC founded The World Industry Council for the Environment (WICE) in 1993 with the objectives of:

- influencing the direction of policy-making towards cost-effective and sound science-based policies in the field of sustainable development;

- being a catalyst for change within industry and demonstrating progress in corporate environmental management.

WICE has since published several documents on industrial environmental issues. Among these is a normalised guideline on the format and presentation of environmental reporting to different audiences [12], directed to those in charge of carrying out environmental audits and introducing environmental management into an organisation. Corporate environmental priorities are set according to the most relevant environmental concerns of each individual firm. In 1995 WICE merged with The Business Council for Sustainable Development (BCSD) to create the World Business Council for Sustainable Development (WBCSD), to continue working within the fields of policy development, business strategy and environmental management, and demonstration/implementation projects, in collaboration with ICC. Membership in WBCSD is by invitation to companies world-wide which seek to provide business leadership in sustainable development and environmental performance.

The ICC Charter and the guidelines published by WICE do not offer any form of external verification or certification other than that of adhering to the ICC

Charter or being a member of WBCSD. Companies which do adhere are encouraged to proceed with auto-declarations of their environmental performance to the public, governments, pressure groups, and other stakeholders. In the WBCSD, the member companies commit to support the Council's work by making their know-how and human resources available. Such a commitment serves in itself as a declaration of environmental intentions.

The position taken on environmental issues by the ICC since 1986 has led to a number of national and international regulations and standards on environmental management, of which BS 7750, EMAS and ISO 14000.

6.3.2 BS 7750

Britain, Ireland and Spain have each adopted legally-binding standards for environmental audits as an instrument for ensuring environmental compliance and improvement. The British Standards Institute's BS 7750 has had the strongest international impact of the three (the others being Ireland's IS 310 and

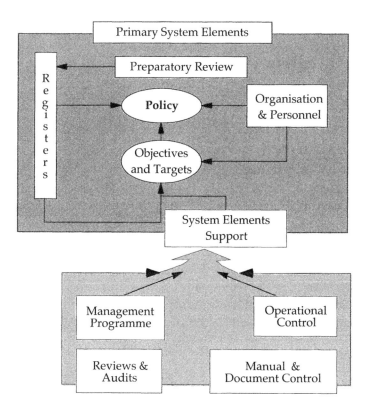

Figure 6.3   Framework of British Standard BS 7750.

Spain's UNE77-801(2)-94). It is designed for compatibility with the Quality Standard BS 5750 so as to allow easier integration into already existing management procedures. As the quality standard it is meant as a qualification of environmental responsibility.

The features of the standard are shown in Figure 6.3. The primary system elements are those who are directly concerned with the implementation of the system into the organisation. The support elements are those elements used for regulating day-to-day activities and generating input for revision of policy, objectives and targets.

The implementation of the system can be divided into three phases. First, a *preparatory review* is made. The review should consider all environmental issues that are relevant to the organisation. Furthermore, it should evaluate the environmental effects of the organisation's activities, evaluate existing environmental management procedures, and review information from occasional environmental incidents.

The preparatory review serves as a base for the determination of the organisational measures which must be taken to maintain a functioning environmental management system. Phase two consists of the actual *implementation of the elements* of the standard (Table 6.1).

Table 6.1   Elements of BS 7750.

• environmental policy	• manual and documentation
• organisation and personnel	• operational control
• environmental effects	• environmental management records
• environmental objectives and targets	• environmental management audits
• environmental management programme	• environmental management reviews

The standard requires a commitment to continuous improvement as a basis for all activities. Evaluation of environmental effects shall not be limited only to those directly related to the company's own activities or products, but will apply to indirect effects such as transport (staff or clients travelling to and from the company site) to the work-place and sub-supplier performance. Furthermore, a register of significant environmental effects directly resulting from company activities has to be kept.

In the third phase, after the initial implementation of the environmental management system elements, compliance to the BS 7750 standard is *verified* by *certified auditors*. The implemented elements of the standard are then subject to regular revision and independent audits of the suitability of policy, objectives, and targets to the environmental status of the company. The performance records resulting from regular audits must be made publicly available.

Certification according to BS 7750 is strictly site-based. Each site must thus have implemented all features of the standard to be certified. This provides a loop-hole, since it is possible to have certified and uncertified plants producing the same products. Although only one of several production facilities for the same product may be registered under the BS 7750, products from other plants may get a "free ride" on this certification. If certification is to serve a purpose, a marking system would have to function to allow the customer to detect whether a certain product came from a certified plant or not.

In addition to establishing structures for environmental management, BS 7750 has also established a verification system, including a register of certified environmental auditors. This is likely to be of competitive importance for companies certified within this system, once an international standard on environmental management is created.

Some critics claim that BS 7750 is not powerful enough, in the sense that it does not sufficiently stress the surveillance of sub-supplier probity. Nevertheless, it is a popular system which is often used as a first step towards registration under The European Community Environmental Management and Auditing Scheme (EMAS). It is also one of the few existing standards on environmental management that can be adapted to the ISO 14000 Series on environmental issues within industry with only small modifications.

6.3.3 THE EEC ENVIRONMENTAL MANAGEMENT AND AUDIT SCHEME (EMAS)

Since 1991, the Council of the European Communities has been developing the Council Regulation Nr 1836/93 [13], called the EEC Environmental Management and Auditing Scheme (EMAS). It came into effect in 1995 with the objective of promoting continuous improvement in the environmental performance of businesses throughout the European Union.

The tools to accomplish this according to the regulation are to:

*   plan and put into practice environmental programmes and environmental management systems at sites;
*   evaluate objectively, systematically and periodically the effectiveness of the implemented tools;
*   provide periodic reporting of environmental performance.

Participation is on a voluntary basis. In many aspects it resembles BS 7750 and other national standards on environmental management. In order to avoid an unnecessary burden on companies wanting to adhere to the scheme, those certified under standards such as BS 7750, ISO 9000, ISO 10011 and CEN/ISO rôle, recognised by the European Commission, are not required to duplicate their procedures in order to meet the corresponding EMAS requirements. In Figure 6.4 the required procedures for registration under the Scheme are displayed.

Each member state of the European Community is responsible for setting up a competent body with the responsibility of registering a site upon request. Once an environmental policy, an environmental programme, and an environmental management system have been established at a site, and once an environmental statement has been submitted to the public and to the competent body, the site can be examined and validated by an accredited environmental verifier. If the site is found to comply with the EMAS Regulation it is added to a national register of participants, which is communicated to the European Commission for publishing in the Official Journal of the European Communities [13]. The environmental verifier is a firm or an organisation which has obtained the accreditation to perform verifications by an impartial institution or organisation designated or created by the member state [13].

As can be seen in Figure 6.4, EMAS registration is site-specific just as for BS 7750, indicating possible loopholes concerning emissions: it is possible to have plenty of registered sites within one company and a few hard polluters. In all, the basic requirements of EMAS are similar to those of BS 7750.

The submission of the validated environmental statement and additional required documents to the competent national body has three legal consequences: the entry of adherents to EMAS in a publicly-available list, public access to the company environmental statement, and the right to publish a statement of participation. The scheme functions on the basis of informal pressure from the public. This means that it is expected that major companies active in the EU will not want to be listed in the public register. This is in itself a potential weakness of the system, since once a company is listed it will be difficult to justify any termination of participation. It may create adverse publicity and encourage authorities to investigate environmental compliance of sites no longer listed. Public pressure differs from country to country, however, and clever selection of sites to register may make registration easier.

The definition of a European regulation on environmental management is the cornerstone of work towards the international standard ISO 14000.

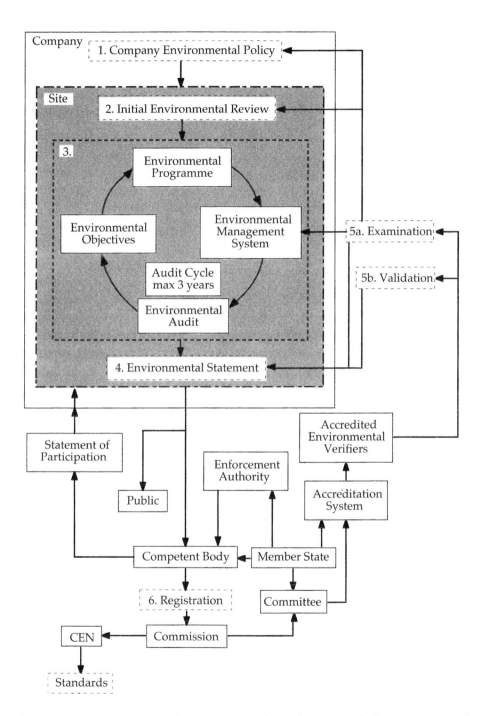

Figure 6.4   An overview of EMAS, © Ruth Hillary, 1993 [14]. Reprinted with permission.

6.3.4 ISO 14000

The International Organisation for Standardisation's ISO 14000 Series has been in preparation since 1991, when the ISO-SAGE (Strategic Advisory Group on the Environment) group was formed as a collaboration between ISO, BCSD (now WBCSD, see section 6.3.1), and the UN, giving birth to the Technical Committee Nr. 207 (TC 207), responsible for normalising environmental management.

With a large number of standards, regulations and guidelines being used concerning environmental issues, there is a need for coherence across borders and markets, to avoid their becoming non-tariff trade barriers. The ISO committee is currently working on a series of standards regulating:

- ISO/TC 207 14001-14009 : environmental management systems (EMS);
- ISO/TC 207 14010-14019 : environmental auditing (EA);
- ISO/TC 207 14020-14029 : environmental labelling (EL);
- ISO/TC 207 14030-14039 : environmental performance evaluation (EPE);
- ISO/TC 207 14040-14049 : life cycle assessment (LCA);
- ISO/TC 207 14050-14059 : environmental terms and definitions (TaD);
- ISO/TC 207 14060-14069 : environmental aspects in product standards (EAPS).

In this section the ISO/DIS 14001 (Draft International Standard) will be discussed, which specifies structural requirements of an environmental management system, and the directional guidelines of ISO/DIS 14004, which support the interpretation of the environmental norms of ISO 14001.

The standard is being developed in line with most of the other guidelines and regulations already in use. In fact, most of the organisations involved in environmental management are also in one way or another involved in the work on the International Standard. This is true for the Environmental Commission of the ICC, as well as for the British Standards Institute, and the European Commission. The aim is to provide a structured environmental management system that assures that a company's performance meets, and will continue to meet, legal requirements and its own policy requirements. An overview of the ISO environmental management system model is given in Figure 6.5.

The development of an ISO standard on environmental management and the EMAS program are the two dominant activities within environmental management in Western Europe. One would expect that if a company was certified under one of these two programmes then it would not be necessary to duplicate already-existing procedures to be certified under the other. This seems to be well provided for in the ISO draft standard. Nevertheless, there is a major difference between ISO 14000 and EMAS: while EMAS is strictly based on individual geographically well-defined sites, the ISO standard is concerned with industrial organisations and operations in their entirety, including subsidiaries and sub-suppliers. It is purposely the intention of ISO to avoid site-based systems.

Figure 6.5   ISO Environmental Management System model.

6.3.5 CORE ELEMENTS IN EMS AND ENVIRONMENTAL REPORTING

The common core components of environmental management systems and environmental reporting are:

- management and system,
- input and output inventory ,
- finance,
- stakeholder relations.

The issues of energy use, water use and discharge, use of chemicals and materials, waste generation and management, pollution abatement and prevention, supplier surveillance and product environmental performance need to be considered in all EMSs, regardless of their detailed structure. These issues will not be discussed in detail in this chapter, since they bear much resemblance to those already discussed in Chapter 4 on life cycle assessment. The same approach to the registration of inputs and outputs is used as in an LCA, this time on a production site instead of only a product or a process. More complete guidelines on environmental reporting can be found in "Guidelines on Environmental Reporting" published by the European Chemical Industry Council (CEFIC) [15] and the environmental reporting guidelines issued by the

Public Environmental Reporting Initiative (PERI) [16] as well as in the standards and regulations discussed earlier in this chapter.

Management and the environmental management system

The information on management and the management system should comprise information on the corporate environmental policies, scope, content, environmental goals and the frequency of their revision.

A description of the company environmental management structure including staff, relationships, responsibilities and procedures of implementation of policies and goals in the organisation should be provided. This should include information on the environmental management system in place, programmes of improvement of environmental performance. Examples of reports providing this are National Westminster Bank Plc 1994, Daimler Benz 1995, Pharmacia 1994, and Baxter 1994.

Input and output inventory

Energy use: the control of energy consumption is a prime target for environmental management. The use of oil, gas, and coal as well as nuclear and hydro energy can be a source of air and water pollution, dangerous solid residuals, depletion of natural resources, acid rain and contributions to the greenhouse effect. Most screening procedures used by financial institutions, banks and insurance companies contain guidelines to avoid companies using energy generated by nuclear power.

Water use and discharge: water has long been considered as a rubbish bin suitable for disposing of waste free of charge. This is no longer true. Regulations in Europe and other parts of the world are becoming increasingly strict. Waste water from industrial operations is disposed of in sewers and treated by water companies who charge for cleaning operations according to the waste content. It is, thus, a cost which can be reduced by increasing the efficiency of plant operations so as to lower output emissions. The positive effect of the management of waste water at the plant is twofold: it lowers remediation costs and increases the availability of clean water.

Use of chemicals and materials: an inventory of the environmentally relevant chemicals and materials used is a primary requirement. These data should preferably be presented in the form of *mass balances* [17].

Several chemicals and materials frequently used within the polymer industry have been identified as hazardous for human health. Among these are PVC, mainly due to of its chlorine content, but also because some its additives, such as stabilisers and pigments, contain heavy metals (Most of these have by now been replaced with other less toxic chemicals).

Environmental criticism attacking the use of PVC because of its chlorine content may be misleading, however; PVC can be considered to be a stable form of chlorine and therefore a good sink for chlorine generated in many synthesis operations.

Further subjects in the spotlight are foaming agents, notably CFCs, which are now forbidden in many countries. Other additives or processing aids under debate, such as brominated and other halogenated flame retardants, cadmium and lead heat stabilisers and release agents which cause damage to the ozone layer, are discussed in more detail in Table 3.5 of Chapter 3.

Wastes: as demonstrated in Chapter 3, waste avoidance and recycling are becoming standard features within the polymer industry, regardless of whether an EMS is in place or not. In some areas, such as the plastic bag industry, in-plant and post-consumer recycling have become prerequisites for commercial survival. The population continues to increase and even if the amount of generated waste per capita remains constant at current levels it is clear that waste generation is a continuously growing problem. Recycling will therefore have to be complemented with measures of waste prevention and reduction of resource consumption for society to approach sustainability. This is, in fact, the general approach of the European Community. Recently, a new directive on packaging has been issued which aims for the withdrawal of 60% of packaging waste in favour of recycling within five years of the date the directive coming into force, and for a 90% recovery rate for packaging in the following 10 years [18]. It is likely that the potential risk of environmental liabilities will increase in the future, further pushing industry to better manage its waste production. A commonly adopted practice is to indicate the nature and volumes of waste generation, ratios of hazardous waste and recycling and methods of treatment and disposal of non-recycled waste. Examples of companies aiming for zero waste in their annual reports are Renault [19] and DuPont de Nemours [20].

Air emissions: the amount and type of emissions, such as greenhouse gasses and ozone-depleting substances, and their corresponding environmental effects should be quantified. Preferably, data should be presented for preceding years to allow progress to be evaluated.

Finance

Environmental spending: the definition of what is environmental cost is not always evident. Pollution abatement costs and other "end-of-pipe" technologies as well as legal compliance costs can certainly be counted as environmental spending, as can environmental taxes and fines. Waste avoidance programs and integrated product and process development towards improved environmental performance incorporate many aspects and it may be difficult to decide just which cost should be allocated to which type of expenditure. Such issues are current subjects of debate within industry itself, so the best strategy for a company is to participate in this discussion in order to be up to date.

Environmental liabilities: liability for accumulated pollution is a greatly contested issue [17]. It can be difficult to estimate the level of liability when no contamination studies have been done or when contamination emanates from several sources.

Stakeholder relations

Employees: The programmes of employee training, and employee awareness and commitment, is important information and must be made available. This information can be generated through surveys on employee attitudes to environmental issues, by disclosing the number of people having gone through training programmes and by internal awards schemes.

Legislators and regulators: this involves keeping up to date on relevant legislative measures for the company, participating in covenants and voluntary agreements on environmental issues, and actively participating in discussions on the development of future legislation within fields of primary importance to the company.

Supplier surveillance: in reality it is impossible to separate the environmental performance of an industrial operator from that of its suppliers, and it is important to address this in the EMS. It would seem rather contradictory to claim good environmental performance while using the services of questionable sub-contractors or sub-suppliers. Companies may be obliged to disclose this kind of information when reporting environmental performance.

This may well call for the introduction of EMSs in smaller companies as well, if they wish to keep their clients. If a smaller company lacks the resources to introduce an EMS it might be obliged to submit itself to the audits of its clients.

Product environmental performance

this subject has been thoroughly discussed in Chapter 5. It is an important element of the EMS in the sense that it facilitates communication with clients concerning the products. The degree to which the company is committed to evaluate the environmental performance of products and processes should be disclosed. Furthermore, design for the environment programmes and the achievement of product life cycle management programmes should be specified.

6.3.6 COST AND UTILITY OF ENVIRONMENTAL MANAGEMENT

At first sight the implementation of environmental management systems would seem to imply enormous costs for minimal gain. Nevertheless, the experience of six sub-suppliers to the Rover motor group in England in the EMAS pilot programme led not only to improved environmental performance but also to quality improvement, cost savings from £ 10,000 to £ 80,000 and new opportunities for energy and waste recycling [14].

A case study carried out at Duffield Printers, a small printing firm performing an environmental review towards registration under BS 7750, reported that as a result of the review, their use of chemicals could be reduced by 50% and some solvents could even be totally eliminated. Additional improvements in aspects of quality and in the cost of lighting led to total savings approaching £ 10,000 per year [21].

Expenditure for compliance with company policies and for environmental reviews, auditing procedures, verification, and registration can vary according to company size. Consultation fees of $ 10,000-50,000 can be expected for small companies, with an addition of $ 10,000-20,000 for intermediate size companies. Results from the UK report costs of £ 15,000 for internal preparation and an additional £ 7000 for external verification [22].

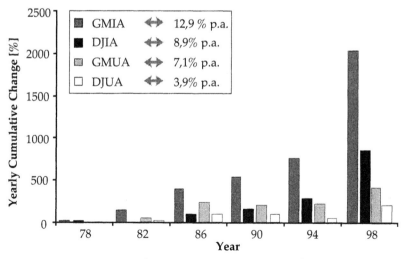

Figure 6.6   Cumulative growth of GMIA and GMUA vs. DJIA and DJUA, 1978-1998 [23].

It is still difficult to predict the relation between costs and benefits of adhering to EMAS. Because of the nature of EMAS, it is important to weigh the benefits of joining against possible losses due to termination of participation. It is recommended that companies wishing to adhere, ensure that they can do so on a long-term basis. The initial cost of implementing stringent environmental management systems might seem prohibitively high in some cases. However, a comparison of the growth rate of traditional stock market indicators such as the Dow Jones Industrial and Utility Averages (DJIA & DJUA) to those of environmentally-screened averages such as the Good Money Industrial and Utility Averages (GMIA & GMUA), as shown in Figure 6.6 would seem to indicate a long-term advantage in managing environmental issues.

The Good Money averages include (as far as is possible) the same number of companies within the same industrial sectors as those in the Dow Jones Averages. This selection takes account of additional criteria of an environmental nature such as for example [10]:

- utilities operating nuclear power plants are excluded from consideration (negative criterion);
- utilities generating power from renewable energy sources are selected (positive criterion);
- companies having implemented energy conservation programmes are selected (positive criterion);
- fossil fuel utilities with particularly strong environmental clean-up records are selected (positive criterion).

As can be seen in Figure 6.6, the additional criteria used in the Good Money Averages have lead to the selection of companies with much higher annual growth. This is the result of the careful screening process aimed at selecting well-managed companies which accord high priority to environmental issues. The mere fact that financial institutions use environmental screening procedures to select the companies in which they invest is no guarantee for success, however. There are several examples of funds which have not done well at all. Their poor performance has been explained by their investing indiscriminately in anything to do with the environment. A majority of the better-performing funds have much more stringent screening procedures, and they tend to focus on alternative energy suppliers, pollution abatement and waste management companies.

It is not necessary to be active only within these sectors to pass the environmental scrutiny of financial institutions, and companies within traditional industrial sectors which effectively manage their environmental performance also perform well. The majority of companies now considered as leaders in environmental change are managing environmental issues using environmental management systems. These systems encourage the user to go beyond compliance to continuously improve performance regardless of existing legislation. To be one step ahead may make all the difference between success and failure when legislative changes affect a business.

6.4 EMS IN SMALL AND MEDIUM SIZED COMPANIES

Up to now primarily multinationals and larger-sized companies have implemented EMS in their organisation. Small and medium-sized enterprises (SMEs) may assume they lack the resources to implement EMS and report their environmental performance to the public. Although this may well be true in some cases, their failure to implement EMS in their organisations would make the European polymer industry rather vulnerable-79% of European polymer companies have less than 50 employees [24]. These companies are vitally important to the economy. They create jobs, are a source of innovation and competition, create a dynamic, healthy market economy and preserve a stable economic base. Investigations in the United Kingdom [21] have shown that despite governmental efforts to inform SMEs of the economic benefits of managing their environmental performance many are still

- unaware of the relevant legislation,
- unconvinced of the potential cost savings and market opportunities,
- out of step with their customers' requirements,
- disassociated from their stakeholders' concerns.

It is not likely that such a large part of a business sector will be able to remain outside of an EMS standard for much longer. Larger companies are beginning to demand evidence of environmental competence from their suppliers, which are in many cases SMEs. Furthermore, as indicated above, financial institutions are realising that environmental performance information is becoming a key feature in the assessment of the sustainability of future debtors' activities. These reasons may be encouragement enough for SMEs to embark upon environmental reporting.

It is thus important that practical standardised environmental programmes be made available to smaller companies without requiring expensive consulting. One possible solution would be for regional development agencies to take on a helping role. Lack of resources should not be a reason for avoiding any sort of environmental management. However small the resources, there can be potential for improving environmental performance. Process analysis is able to identify unnecessary waste generation or energy use that can be remedied by a simple change of operating procedures. At the simplest level, this could mean turning off the heating in rooms that are not frequently used, or closing windows. The UNEP Technical Report Nr 24 [17] reports the following minimum number of issues to be addressed in environmental reporting in smaller companies with limited resources:

- the company's latest environmental policy statement, with dates of any reviews;
- a description of the company's environmental management systems;
- an outline of management responsibilities and reporting links for environmental protection;
- an account of the company's legal compliance record.

Concerning inputs and outputs to the industrial process the following minimum elements are recommended:

- materials use and trends,
- energy consumption and trends,
- water consumption and trends,
- health and safety,
- environmental accidents,
- major waste streams,
- air emissions,
- water effluents,
- product impacts during use.

Financially, there are two recommended priorities:

- the level of environmental expenditure,
- the extent of the company's environment-related liabilities.

Lastly, the company should demonstrate how it is working with five key stakeholders:

- its own employees,
- government legislators and regulators,
- the local communities near to the company's facilities,
- the company's shareholders and investors
- the company's involvement in any relevant initiatives launched by business and industry associations.

A closer glance at EMAS shows that the problems of SMEs have been taken into consideration in drafting the regulation. The importance of participation of small and medium-sized companies is stressed. Their involvement is to be promoted by establishing or promoting technical expertise and support. Extensive reporting such as that generated by multinational corporations is obviously not possible for smaller companies.

Environmental auditing is only one feature of a much larger system of measures to improve environmental performance. Large manufacturing facilities will generally demand more detailed and formal EMSs than smaller facilities.

6.5 TOWARDS GREENER MANAGEMENT. A SUMMARY

Technical solutions to environmental problems are being developed. Their application to real industrial structures requires a high degree of organisation. If environmental considerations are to be efficiently addressed they must be systematised. This is being addressed in several parts of the world, and most multinational and larger companies have already implemented some kind of environmental management system.

The implementation of such systems is not free from cost, and they demand a certain level of commitment from their participants. There are nevertheless inherent advantages of managing environmental issues in a systematised way, in terms of access to finance and insurance, but more importantly to identify potential liabilities and avoid unnecessary expenditure on clean-up and non-compliance. Other reasons for adhesion to a standardised EMS is the possibility of gaining competitive advantage in international trade. Non-adherence could actually become a trade barrier, hindering non-adherent companies from access to key markets. European customers are already requesting conformity to environmental standards in bid specifications for products and services coming from outside Europe [25]. Possibly adherence could also be promoted by giving co-operating companies preference in competitive bidding for public contracts, procurement and preference in investment for adhering foreign investors.

Most of the EMS standards and regulations provide support for implementation in small or medium-sized companies. These companies can expect that larger clients will submit their environmental performance to scrutiny. In order to retain contracts, having implemented environmental management and having a record of clear environmental policies related to the business may be a competitive advantage. The provision of professional assistance in the implementation of environmental management systems adapted to SMEs will certainly lead to more of them embarking upon environmental management by their own choice, rather than being forced to do so by the market.

The creation of national and international standards for environmental management systems is no guarantee for environmental improvement. Neither does the existence of an EMS within a company make its activities environmentally sound. The increasing adoption of environmental management systems as a core management tool will probably create a common language between companies to aid quicker learning and progress.

REFERENCES

1. J. R. Stock and D. M. Lambert, *Strategic Logistics Management*, 2nd ed., Irwin: Homewood, USA (1987).

2. B. Bergendahl, *Environmental Regulation in Sweden* in *Regulating the European Environment*, T. Handler Ed., Baker & McKenzie: London, pp. 147-154 (1993).

3. P. L. Hauck and R. A. Smith, *Integrated Waste Management* in proceedings of *Dornbirn '95* (1995).

4. *Polymer Recovery*, GE Plastics / Ravago Plastics (1991).

5. *Environmentally-compatible car scrapping*, Volvo Car Corporation, Report 47.

6. *Svensk Miljöfond: Det Naturliga Stegets Miljöratingmodell*, Det Naturliga Steget Miljöinstitut AB, Amiralitetshuset, Skeppsholmen, S-111 49 Stockholm, Sweden, September (1994).

7. *GEO BANK: The natural way of banking*, GEO BANK, 26, rue Adrien-Lachenal, Case postale 1211, Geneva, Switzerland.

8. I. Jenkins, *Chemical, Waste Stocks Pose Dilemmas for Europe's "Green" Funds*, *International Herald Tribune*, p. 15 (1994).

9. A. Sullivan, *Looking for Shares That Meet Your Ethical Criteria?*, International Herald Tribune, August 27-28, p. 15 (1994).

10. M. Brenner, *Qualifying companies for receiving "green money" for business expansion in environmental industrial energy projects* in proceedings of *GLOBEC '96, 9th Global Environment Technology Congress*, Davos, Switzerland, pp. 9.4.1-9.4.8 (1996).

11. *The Business Charter for Sustainable Development, Principles for Environmental Management and "Model Questions and Answers"*, The Environmental Commission, International Chamber of Commerce, Paris, France (1991).

12. *Environmental Reporting: A Manager's Guide*, World Industry Council for the Environment (WICE), Paris, France (1994).

13. Council Regulation, Nr 1836/93, Art. 2, EEC, 29 June (1993).

14. R. Hillary, ICCET, 48 Prince's Gdns, London, UK: Personal Communication.

15. *CEFIC Guidelines in Environmental Reporting for the European Chemical Industry*, European Chemical Industry Council Brussels, Belgium (1993).

16. *PERI Guidelines*, Public Environmental Reporting Initiative, USA (1993).

17. *Company Environmental Reporting: A Measure of the Progress of Business and Industry Towards Sustainable Development*, UNEP/SustainAbility, Ltd, Technical report Nr 24 (1993).

18. J. Scherer, *Environmental regulation of the European Community* in *Regulating the European Environment*, T. Handler Ed., Baker & McKenzie: London, pp. 1-21 (1993).

19. Renault, *Target: Zero Waste: Automobiles*, The Research & Development Collection.

20. DuPont, *Environmentalisme d'entreprise: Bilan Europe 1993*, DuPont de Nemours International S.A., Environmental Affairs-Europe, Le Grand-Saconnex, Switzerland, 1 (1993).

21. *Small Firms and the Environment Status Report*, Groundwork National Office, 85 Cornwall Street, Birmingham B3 3BY, UK (1996).

22. J. Scherer, Baker & McKenzie, Döser Amereller Noack, Bethmannstraße 50-54, 60311 Frankfurt am Main, Germany: Personal Communication.

23. R. P. Lowry, *GOOD MONEY: A Guide to Profitable Social Investing in the '90s*, W.W. Norton & Company: New York (1993).

24. O. Krugloff, *Plaster: Materialteknisk Handbok* , Svenska Återvinningsföreningen:, p. 10 (1995).

25. M. Blazek and B. Dambach, *Integrating ISO 14000 Environmental Management Systems and Design for Environment at AT&T* in proceedings of *IEEE International Symposium on Electronics and the Environment*, Dallas, USA, IEEE, Piscataway, NJ, USA, pp. 255-259 (1996).

7

CASE STUDIES

Two studies of pharmaceutical packaging and networking within the automotive industry are presented to illustrate how environmental requirements affect industrial activities. Directives on producer responsibility are forcing OEMs to organise the recovery and disposal of their products with the lowest possible environmental intervention. This pressure is translated down through the supplier's chain to sub-suppliers, material producers and end-of-life industries. Requirements on recycling and environmental performance improvements demand organisational and technology change in order to retain performance and profit.

7.1 PACKAGING RESTRUCTURING

Plastic packaging is a controversial subject in the environmental debate. On the one hand, plastics compare favourably with other materials, as they lower fuel consumption in transport due to higher transport capacity per unit weight. In addition, plastic packaging is easily collapsible when emptied, which facilitates convenient waste handling. On the other hand, management of packaging waste can still be problematic. In this case study we shall see how an environmental strategy is disseminated from corporate level into a business sector of a medical company and how this has affected the sectors' packaging strategy.

7.1.1 PHARMACIA & UPJOHN

Pharmacia & Upjohn Inc. is one of the fifteen largest pharmaceutical companies in the world. It was formed in 1995 through a merger of the Swedish company Pharmacia AB and The Upjohn Company from the US. Pharmacia & Upjohn delivers products in several therapeutic areas such as oncology, growth and metabolic disorders, neurology, nutrient administration and smoking cessation.

The situation of the pharmaceutical industry is somewhat different from that facing many other industries: their products are aimed at improving the health of their consumers. Consequently, it would be rather ambiguous to maintain this aim without considering the environmental impact of company activities as a whole. Pharmacia & Upjohn has implemented an environmental and safety management system, based on ISO 14000, in order to fulfil its commitment to reduce resource consumption and environmental interventions caused by its activities. The introduction of the management system has brought on a number of changes in the organisation.

"The programme has a strategic value. It will keep the company more aware of its own status, not only from an environmental point of view, but also in terms of management control", says Gösta Larsson, pioneer in the development and implementation of the environmental management system in the former Pharmacia company. *"Keeping attention on what is happening on the international scene is also profitable, since this will avoid unpleasant surprises in terms of legislative changes."*

Pharmacia & Upjohn, Nutrition

Pharmacia & Upjohn, Nutrition, is a world-leading manufacturer of parenteral nutrition solutions, such as fat emulsions and amino acids and intravenous (I.V.) solutions (a general term for glucose and salt solutions). Historically, a multitude of packaging solutions has been used within the different product ranges. Glass bottles dominate the nutrition field and plastic packaging the field of I.V. solutions; in the latter case the principal plastic packaging material has been PVC.

7.1.2 PACKAGING DEVELOPMENT

In line with the environmental programme initiated at corporate level, a packaging restructuring strategy was developed within Pharmacia & Upjohn, Nutrition. It was decided that products should be provided in the safest, most convenient and environmentally-adapted packaging possible. The multitude of packaging solutions would be replaced by a family of packages manufactured using a minimal amount of (preferably) a single material. A harmonisation of material use would ultimately allow disposal of the entire product range into the same waste flow. The restructuring of the packaging strategy for nutrition solutions and I.V. solutions is the focus of this study.

In order to select between glass or plastic packaging, a life cycle inventory (LCI) was carried out [1]. Polypropylene was tentatively selected as the most suitable polymer, based on the range of grades available, from rigid (for bottle manufacturing) to flexible (for bag manufacturing).

Two packaging alternatives were studied. The production units (see Chapter 4) were:

1)   A 500 ml glass bottle with a rubber stopper and an aluminium protective capsule.

2) A 500 ml polypropylene plastic bag with a port system comprising a rubber stopper. This bag consists of an inner bag wrapped with an overpouch, the composition of which depends on the content of the inner bag. For an oxidation-sensitive solution (fat emulsions and certain amino acids), the overpouch contains a high-barrier layer, but will still consist primarily of polypropylene.

Due to legislation, primary medical packaging must be made from virgin materials. Material flows in the LCI were therefore based on virgin raw materials. As a simplification, the same holds true for secondary packaging and pallets. Furthermore, it was assumed that all material would be incinerated after use. In reality recycling schemes might be set up, as will be seen later in this chapter. Energy consumption and emissions at filling and sterilisation were assumed to be equal for the two alternatives.

The study showed that the total energy consumption and the total emissions would be lower for the plastic bag than for the glass bottle. Transports constitutes a large part of the environmental impact; the greater environmental impact associated with the glass bottle is thus a consequence of its higher weight.

The same application requirements as for all existing products would have to be satisfied by the chosen plastic. Thus the polymer material would need to have (or allow modification to create):

• resistance to lipid solutions,

• resistance to amino acids,

• resistance to final sterilisation,

• transparency (even after sterilisation),

• barrier functions to prevent oxygen diffusion.

Further requirements are that the material should be suited for mechanical as well as energy recycling, to meet regulations in different countries. The proof of existing recycling technology and secondary applications for the reprocessed material would also be necessary to ensure environmental credibility. Several points would need to be fulfilled for an efficient recycling chain:

• suitable design,

• collection,

• cleaning,

• recycling technology,

• characterisation,

• secondary applications, which does not include reuse for primary medical packaging due to safety regulations, as mentioned previously.

The selected packaging, the Excel® bag, is shown in Figure 7.1. The Excel® material is a polypropylene (PP)-based multilayer film, including a polyester ether outer layer and a styrene-ethylene-butylene-styrene (SEBS) tie layer. A natural rubber (NR) stopper is present at the port system of the Excel® bag.

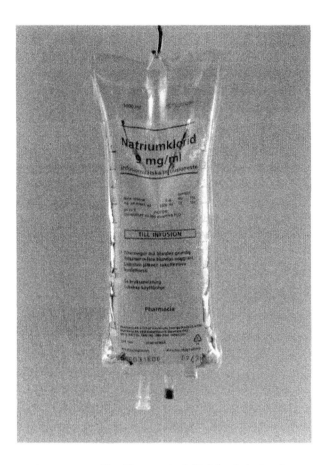

Figure 7.1   The Pharmacia & Upjohn infusion bag.

7.1.3 PACKAGING RECYCLING

To evaluate the attitudes of users towards the change of packaging material and to evaluate the efficiency of material sorting and the properties of the recycled material, Pharmacia & Upjohn performed collection and sorting trials with I.V.-solution bags at wards of six Swedish hospitals. The tests were performed during Spring and Summer 1995, when the Excel® bag and the old PVC bag were both present on the market.

Nurses were instructed to separate, as completely as possible, the polypropylene-based Excel® bag from the NR stopper and from other types of infusion containers, mainly made from PVC. Administration sets, mainly consisting of PVC, were also removed from the bag prior to disposal in the waste container. The waste containers were collected by a forwarding agency and brought to Pharmacia & Upjohn where they were checked for mistakes in the sorting procedures. Samples were then taken for material characterisation and for manufacturing trials. Two possible secondary applications were identified: boxes for waste collection and flower pots.

Sorting and collection

The different types of "contaminations" resulting from incomplete sorting will affect recycling in different ways. The natural rubber stoppers cannot be removed in the reprocessing step, and will have a degenerative effect on mechanical properties of the polymer. Paper tissues and similar contaminants are, on the other hand, relatively easily removed in a washing/cleaning step prior to regrinding. PVC presents a problem due to the formation of hydrochloric acid, HCl, and it is important to identify the limit of PVC contamination which the process can accept. A limit of 1 % has been indicated [2]. Within the three-month period of the test, there was a significant improvement in the quality of sorting, and although the level of contaminants in some few waste containers exceeded allowable limits, as a whole the sorting was good enough to enable reprocessing.

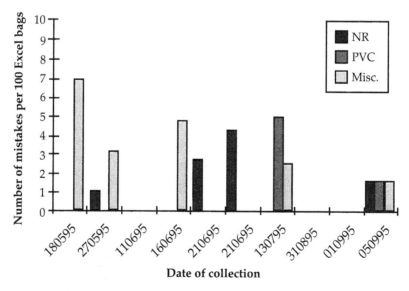

Figure 7.2   Collection results from one hospital ward.

Figure 7.2 is an example of the results from the sorting of I.V. solution packaging waste from one of the wards. The data is from sorting at different dates during the test period. Included are the number of NR-stoppers, PVC bags (including administration sets as well, since these are predominantly made of PVC) and miscellaneous contamination found in 100 correctly-sorted Excel® bags. Miscellaneous contamination was predominantly paper tissues but also contained other types of bags (such as polyethylene or aluminium), syringes and other waste.

The transportation of waste bags from the hospitals to Pharmacia & Upjohn was identified as the most cost-intensive stage of the recycling process. One major reason for this was that insufficient infusion solution packaging waste was collected to regularly fill the collection vehicles. It would therefore be important to collaborate with manufactures of other sources of polypropylene waste within the hospitals in order to increase the collectable quantities. It was also recommended that local and/or regional material recovery facilities should be identified and used, to limit transport distances. In addition to raising costs, transportation probably has the largest environmental impact of the steps in the recycling process.

Reprocessing and characterisation

Pharmacia & Upjohn has embarked on academic co-operation with the Swiss Federal Institute of Technology in Lausanne (EPFL) to develop methods for recycling and characterisation as well as to identify possible secondary applications for re-processed materials.

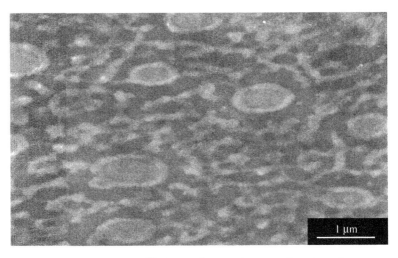

Figure 7.3   TEM image of Excel® blend where polyester ether domains are separated from the PP/SEBS matrix by SEBS which acts as a surfactant.

The reprocessing of Excel® bags was found to give a compatible and partially miscible blend as shown in Figure 7.3. The figure depicts the three-phase microstructure observed by transmission electron microscopy (TEM). Thermal analysis identified the glass transitions of PP (-10 °C) and of the polyester ether (55 °C), and the melting of the PP phase (130 °C). Such a low melting temperature is indicative of the miscibility of the blend [3].The elastic modulus of the recycled blend, was too low for applications such as collect boxes. Several blends were prepared by EPFL in order to increase the stiffness of the material. Examples include blends with low (LM) and high (HM) moduli polypropylenes and with polyamide (PA 6). Several talc-filled blends were also processed, including a PP-talc masterbatch with the aim of increasing stiffness at low cost.

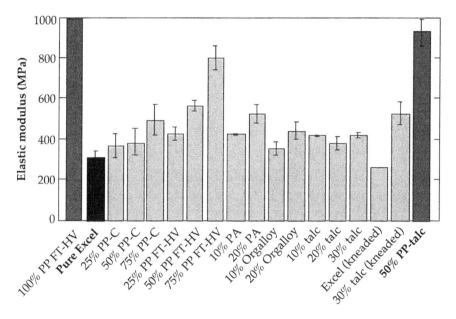

Figure 7.4   Results from Excel® blending tests.

As shown in Figure 7.4, blending the Excel® material with a 50% (by weight) of PP-talc masterbatch provides the required performance, similar to virgin high modulus PP, for the production of stiff products. The mechanical performance of the recycled material was found not to be degraded by PVC fractions of up to of 1% by weight. It was slightly improved by the presence of PP labels, and insensitive to the presence of ink. The long-term mechanical performance of the talc-filled recycled material was found to increase by 15-20% due to inhibited PP recrystallisation caused by the complex blend morphology. As long as the material is used at temperatures below its 55°C glass-transition, oxidative degradation can be considered to be negligible.

Recycled plastic material of this composition was used for the manufacture of flower pots at Hackman Household AB in Gislaved, Sweden, a company which had previously shown great interest in receiving recycled-polypropylene compatible materials. The collaboration between Pharmacia & Upjohn and Hackman shows the importance of finding fruitful collaborative ventures in order to succeed in recycling processes.

7.1.4 ATTITUDES OF HOSPITAL PERSONNEL

Simultaneous to the recycling study, Pharmacia & Upjohn presented a questionnaire to the hospital personnel regarding their attitudes on environmental issues and plastic materials. The questionnaire was presented twice, prior to and after the case study, in order to investigate if attitudes changed during the study.

The questionnaire showed a strong environmental commitment among the personnel, which was strengthened by the study. No indication of lack of time for source separation or any lack of top management commitment was found. The reactions to the study were very positive. This was confirmed by the fact that the hospital wards actively proceeded with source separation of their packaging waste after the study.

"The attitudes to Pharmacia & Upjohn became more positive during the study", says Bengt Mattson, Pharmacia & Upjohn - Corporate Environment & Safety. *"This shows that making environmental commitments gives serious pay-back in terms of good-will. The very close collaboration with the hospital personnel also helped us to identify possible improvements in our products and systems. Another important conclusion from the study was that the nurses felt they lack information and knowledge on environmental issues. This should therefore be one of our contributions in these collaborative efforts."*

Figure 7.5 gives the answers to questions regarding plastic materials. As expected, PVC is regarded as an environmentally hazardous material. The negative feelings connected with PVC are most probably the explanation for the generally negative attitudes towards plastics. If PP (polypropylene), PE (polyethylene) and PET (polyethylene terephthalate) are considered, the attitudes are slightly more positive. It is obvious that PP is perceived as more environmentally sound than PE. The clear difference in attitudes towards PP and PE, already in q1, was slightly unexpected. The study has made this difference even stronger. This is explained as an increase in knowledge regarding successful PP recycling. As expected from the beginning, PET was considered the most environmentally appropriate plastic material. This is, of course, due to the refilling and recycling of PET bottles for soft drinks.

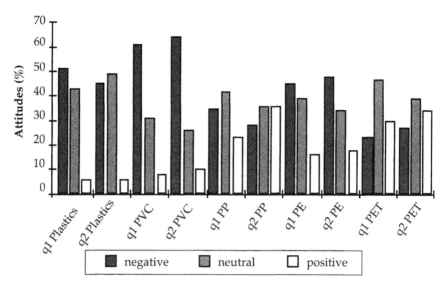

Figure 7.5   Results from questionnaires (q1: before the study, q2: after the study).

7.1.5 THE VITAL RECYCLING CHAIN

Recycling of packaging waste is becoming mandatory in an increasing number of countries all over the globe. The public pressure on industry to face up to its environmental responsibilities offers potential marketing benefits for pro-active companies. The Pharmacia & Upjohn case shows that it is possible to recycle I.V.-solution plastic packaging by starting with sorting at the source of waste generation. Knowledge about polymer miscibility, blending and long-term performance is necessary in order to obtain an appropriate secondary material.

Transportation is a possible obstacle for future recycling schemes. Waste transport might be too expensive if the amount of sorted waste is not sufficient to fill trucks for each transport run. Transportation also has a heavy environmental impact.

A survey of the attitudes of hospital personnel indicates that Pharmacia & Upjohn have gained good-will through their recycling efforts. This enables fruitful discussions between the producer and the customer on environmental improvements on product and packaging designs.

Close collaboration with the recycling company and the user of the recycled material is essential. The Pharmacia & Upjohn case study shows clearly the importance of fulfilling all the requirements of the vital recycling chain.

7.2 INDUSTRIAL NETWORKING FOR COMPETITIVENESS

The automotive industry is making increasing use of polymers and polymer-based composites. This case study concerns the recycling of an automotive component made from a polymer composite.

The radiator end-cap of an automotive radiator sub-assembly is currently made from glass fibre-reinforced polyamide 66 (PA 66+GF), which is particularly suitable for under-the-bonnet applications as it maintains high stiffness and strength at high temperatures. For the radiators mounted in cars made by the Volvo Car Corporation, the raw material is supplied by DuPont de Nemours to the component producer, AB Konstruktions-Bakelit. This company then supplies the finished component to the sub-assembly producer which then ships the finished assembly to Volvo.

Before studying in detail the operations necessary for end-cap recycling, the current situation within the automotive industry regarding recycling will be discussed. The environmental policy of each of the companies concerned will be presented.

7.2.1 PLASTICS IN THE AUTOMOTIVE INDUSTRY

The automotive industry is increasingly being singled out as a target for environmental pressure. It is estimated that 500 million cars are in operation today. An additional 50 million cars are produced annually, while 6 million are discarded annually in Western Europe [4]. The industry has a strong impact on the environment, not only because of large amounts of waste generated from end-of-life vehicles (elv) (Figure 7.6), but also because of the atmospheric pollution produced by the vehicles during service. Occasionally, drastic measures are taken to limit traffic pollution by regulating traffic or by adopting non-emission regulation for vehicles circulating within city limits [5].

Plastics offer several advantages over traditional materials in automotive applications, namely:

- manufacturing process cost savings, lower tooling costs, reduced number of parts, reduced assembly line costs;
- design freedom, allowing functional integration and material streamlining;
- safety enhancement by means of more resilient body shell components;
- high performance to weight ratio, allowing optimised light weight design;
- corrosion resistance;
- noise reduction.

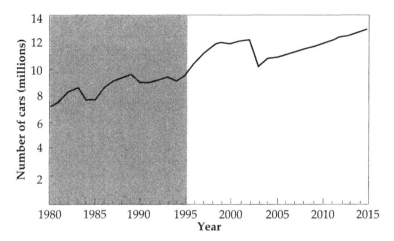

Figure 7.6 End-of-life vehicles per annum in Western Europe up to 1995 and estimated to 2015 (the estimations assume a life of ten years and exports to the eastern block and third world countries of 10% of the recorded end-of-life vehicles) [6].

Automotive recycling directly brings the focus on plastics. The composition of an average car is shown in Figure 7.7. Cars have long been dismantled to recover reusable or valuable parts such as radiators, bumpers, batteries and fuel tanks before being shredded to recover the metal content.

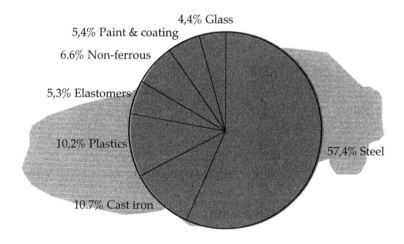

Figure 7.7 The material composition of an average car [12].

More than 90% of the total number of used cars are shredded for recovery and disposal of their material constituents [7]. Each recycled automobile leaves around a hundred kilograms of Automotive Shredder Residue (ASR). This residue is a complex mixture of plastics, wood, paper, tar, road dirt, glass, metals, water, and various automotive oils and fluids (Figure 7.8). There is currently no available economic method to separate more than a limited amount of the constituents for recycling, and most ASR is landfilled. Yet space for landfill is becoming scarce and expensive. Legislation in Germany classifies ASR as hazardous waste, making it four to five times more expensive to landfill. The promised revision of the list of "special wastes" in France could make the cost of dumping around 40 times higher [8-10].

> **"The recycling problem in a word: Plastics"**
>
> "Car Trouble" World Resources Institute [11]

The plastics share of ASR is increasing rapidly as auto makers are lowering vehicle weight to improve fuel efficiency during service. About 700 different grades of plastic resins can be found in a vehicle [7], and with current design practices most of these plastics will end up in ASR from which only a small proportion can be recovered for recycling. In the United States alone around 200 shredders produce 50,000 metric tons of ASR annually out of which only 30-40% of a limited number of plastic parts such as bumpers are separated and reused [7]. Clearly this is a problem, but it is also a business opportunity as land-fill costs increase. Car producers will have to include recyclability among the design criteria of its products.

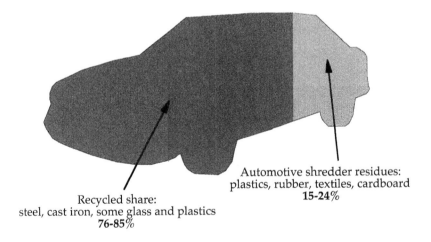

Automotive shredder residues:
plastics, rubber, textiles, cardboard
15-24%

Recycled share:
steel, cast iron, some glass and plastics
76-85%

Figure 7.8   Recycled and ASR shares of scrapped cars.

In many countries automotive producers are voluntary taking on or being imposed by legislation the responsibility for recycling cars. Recommendations from the DGIII EOL Strategy Group of the European Commission propose that 85% of retired vehicles be recovered or recycled by the year 2000, and attach great importance to the implementation of recycling schemes for plastics. These recommendations go even further by suggesting that 95% of vehicle weight be recovered and recycled by the year 2015 [13].

Since the use of plastics is ever increasing in automotive applications the need for effective plastics recycling will further increase in the future. The pressure on OEMs to recover and recycle their products will be felt throughout the production chain by recyclers, dismantlers, shredders, sub-suppliers, contractors and raw material producers. There is thus a strong motivation for the automotive industry to learn to manage its products and processes in an environmentally-efficient way.

Networking within the automotive chain

Automotive OEMs are already accepting the responsibility of setting up functioning recycling networks for their products. In some countries this is an initiative taken by the OEM itself, whereas in others it is by mutual agreement with authorities, while yet other countries are forcing industry to do so by legislative means. Efforts to improve environmental performance during the life cycle of autos gives new responsibilities and new functions to the members of the automotive production chain. The effect of environmental legislation and of the increased focus on recycling within the automotive industry on an OEM (Volvo Car Corporation), a sub-supplier (AB Konstruktions-Bakelit), and a material supplier (DuPont de Nemours International S.A.) will be analysed.

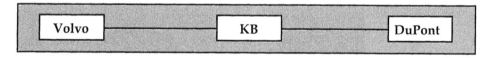

The Volvo Car Corporation (VCC) is a small specialised car producer which has long been exposed to changes in consumer preferences and legislation. Volvo has chosen a proactive approach to environmental issues. As part of its environmental policy, the company frequently takes part in discussions with authorities, politicians, schools, competitors and environmental organisations, the goal of which is to establish co-operation and feedback on its activities.

AB Konstruktions-Bakelit (KB) is a medium-sized company with 400-500 employees. It is mainly involved in business-to-business activities with the automotive industry as the main client. This keeps the company away from direct exposure to public environmental pressure. For KB, the main driving forces for improving environmental performance are the environmental pressure transmitted from OEMs to sub-suppliers and the legislative pressure that effects all industry. For this reason there is no specific environmental activity within the company, more than that reflected in the satisfaction of customer demands and legislative compliance. Instead, KB follows the development of the debate, building knowledge in its core activities of design, processing and characterisation.

DuPont de Nemours has established itself as a major supplier of engineering thermoplastics in Europe. As environmental considerations become increasingly important, DuPont is aiming to place itself as a leader of environmental change. In 1988, the chairman, Edgar Woolard, called for "corporate environmentalism" in industry. This he defined as *"an attitude and performance commitment that places corporate environmental stewardship fully in line with public desires and expectations."* A company aiming at overall excellence must show that it is prepared to go beyond compliance and participate in the forefront of development.

7.2.2 VOLVO CAR CORPORATION

As a specialised car manufacturer Volvo has a strong reputation for safety and quality. The surge of public environmental awareness makes the inclusion of environmental considerations into the fundamental values of the company a logical step, and Volvo has the ambition to become a leader in this area. This orientation adds several issues to those traditionally considered in car manufacture, as well as new organisational requirements, as indicated in Figure 7.9.

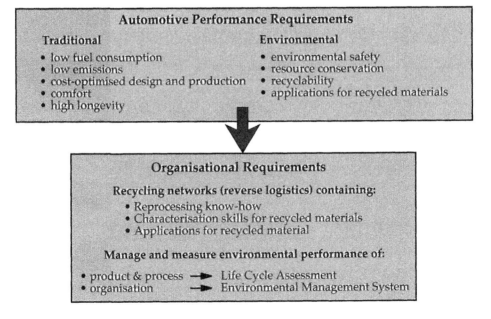

Figure 7.9 Environment-related product performance and organisational requirements for the automotive industry.

Environmental aspects of product design have been considered within engine construction for several years, and various groups have been working with other environmental aspects of auto production and design throughout the company. In 1992 the Department of Weight Reduction and the Environmental Department merged, resulting in the creation of the VCC Competence Centre Environment (CCE) in 1994. The aim of this organisational change was to better co-ordinate the work within product development by combining the responsibility of specification and production in one unit.

Ulla-Britt Fräjdin Hellkvist, head of the CCE states that *"The primary task is to make visible the work that is being done to improve company environmental performance on the product and process level"*. Furthermore, the CCE should:

- take initiatives for process and product improvement,

- co-ordinate and support already-ongoing projects,

- follow up goals and strategies set on the corporate level,

- serve as an environmental messenger and as a meeting place for environmental questions within the corporation.

A life cycle assessment tool (described below) is being introduced into the product development organisation as an aid in product development.

Volvo early realised that there lay a danger in providing information to the public on environmental issues without taking an active part in the environmental debate. By joining the debate, mis-interpretations that could unduly harm the company's reputation can be avoided. Thus, seminars are held for journalists and authorities on environmental issues to foster discussions across societal borders and to show that all parts of society are actors on the environmental scene. Co-operation between universities, industry, environmental organisations and customers provides for faster changes than if each party were to work independently. *"Volvo is showing the public where the company is in its development process and is seeking feedback from the public and industry"*, Mrs Fräjdin Hellqvist continues. *"This puts further pressure on the company to achieve results. It also puts Volvo in the driver's seat in the environmental debate."*

In addition to the CCE, Volvo is using the Volvo Environmental Management System (VEMS) to ensure that its work on environmental improvement is effective. *"Volvo has a long tradition of environmental management systems"*, says Inge Horkeby, environmental accountant of the Volvo Corporation. EMS systems are by no means new to the industry; on the contrary, they are attempts to systematise and standardise already-existing structures for the management of environmental issues. Volvo has an internal environmental management and auditing system which has been in operation for several years. All Volvo plants world-wide have been subjected to environmental reviews. The current system is individually linked to each plant's demand on environmental measures, be they due to legislation or to internal environmental goals of the company itself. *"It is likely that the system satisfies BS 7750 on a general scale, but this alone is no reason for Volvo to have itself BS 7750-approved,* Mr Horkeby continues. *"The company can continue on working on its own system until an international standard such as ISO 14000 is established".*

"BS 7750 has the disadvantage of not having a clear connection to production", Mr Horkeby claims. This makes it somewhat unattractive to Volvo. The EMAS structure, which is a voluntary reporting scheme, is better suited for Volvo from this point of view. It has stronger links to production and sub-supplier surveillance which is an important concern for Volvo. Although it is not yet a requirement, having an implemented EMS is considered as a guarantee of quality, and Volvo would like all sub-suppliers to operate one. Volvo expects the introduction to bring more rigour and clarity to the way in which environmental questions are addressed within a company. It may possibly give advantage in negotiations for contracts with the public sector.

How, then, is the Volvo policy on the environment reflected in the activities of the company?

Product development: safety & recyclability vs. low weight

Traditional design criteria are weighed against environmental considerations to find optimal solutions for future design solutions. The pressure is strong on car manufacturers to use more recycled materials and to design their products for easier recycling. Nevertheless, the environmental savings of such measures are marginal compared to the energy consumption during the service life of a car. The most significant environmental improvements can be achieved by developing low-emission engines with catalytic converters, changing the engine concept to electric or hybrid engines (such as the one adapted in the Environmental Concept Car (ECC) project and in the Volvo 850 Bi-Fuel model running on petrol or natural gas) and by reducing fuel consumption through lightweight design.

In line with the target of the Environmental Car Recycling in Scandinavia project (ECRIS) to produce input on design for recycling and serviceability for design engineers, Volvo has introduced a directive for design for weight savings, recyclability, and serviceability, including time limits for disassembly. This directive is communicated to sub-suppliers. It features items such as functional integration to ease quick disassembly and to streamline the material composition of the cars. To further improve material separation and identification all pure plastic parts weighing over 100g in cars produced after 1991 (that is, about 80% of plastics) are marked for visual identification.

Material selection

Volvo has developed a material selection strategy in close co-operation with its sub-suppliers. *"The current preference is to choose clean and simple materials"*, says Carl-Otto Nevén, representing Volvo in the Product Ecology Project [14]. This is a project co-ordinated by the Federation of Swedish Industries, aimed at developing user-friendly PC-based tools for calculating the total environmental

VCC
Volvo Environmental Concept Car

An effort to adapt car production to calls for environmental awareness within the automotive industry.

Aim
To stay within realistic design boundaries for bringing the car on the market at the turn of the 21st century.

To show the public where the company is in its development process and seek feedback from the public and from industry

Environmental considerations in production, service and final disposal

- inner panels built for easy disassembly without metal parts or material mixtures that impede recycling

- hybrid engine running on gasoline, other fuels or electricity

- integrated front bumper and grille for weight and energy saving and for promotion of recycling by parts consolidation

- six kilogrammes of recycled plastics:
  PP: wheel housing, bumpers, panels
  GMT: floor of luggage compartment
  ABS: ventilation housing

impact of products from the "cradle to the grave". For polymers, this implies choosing higher performance plastics to satisfy the functional requirements and prolonging service life. The residual value and performance of engineering thermoplastics at end-of-life allows for more economically-viable recycling than for commodity plastics. In addition, cascade applications such as light housings and under-the-bonnet components can be found for which cosmetic features are of lower importance.

Product life extension

One often-cited way to improve environmental performance is to prolong the service life of the vehicle [15]. In auto production, however, there is a risk that such measures could be counter-productive. An average service life exceeding 20 years would significantly affect technology renewal. Another possibility is to design cars for upgradability, that is to enable progressive replacement of old technology components without changing or scrapping the car. Volvo maintains that the optimum solution is to design cars for a useful lifetime of 15 to 20 years and to set up efficient reclamation networks. Such a lifetime is long enough to allow substantial technological progress to take place without generating an unacceptably high turnover of cars. It is not very likely that parts in cars with a lifetime of 15 to 20 years will be reusable, since designs change quite frequently for reasons of competitiveness, performance, and safety.

Recycling

According to an official agreement between the largest companies within certain sectors of Swedish industry and in line with proposed legislation [16], these companies shall be responsible for setting up the necessary structures for recycling within their branch of industry [17]. Each producer can choose whether to be directly involved in the recycling of his product or not. Automobile manufacturers for example, agree that there is no reason to enter into car recycling when there is already a functioning network of car dismantlers; they will instead make sure that the largest operators are developed into car recyclers. The manufacturers will, on the other hand, provide all necessary information to the recyclers to enable fast and energy-efficient dismantling, materials identification and separation.

A dismantling plant set up in 1994 within the ECRIS project is evidence of the agreement to create a recycling structure in Scandinavia (Figure 7.10). Volvo co-operates with the shredding companies Stena Bilfragmentering AB and AB Gotthard Nilsson. Together with a third partner, the dismantler JB Bildemontering AB, they aim to create a model for Scandinavian automotive recycling [18].

The ECRIS project has four main objectives:

- to develop methods for the effective dismantling and sorting of environmentally-harmful components and recyclable materials;
- to calculate the environmental load imposed by different recycling alternatives using life cycle analyses;
- to evaluate material recycling and energy recovery methods in practical sub-projects;
- to promote and evaluate markets for recycled material.

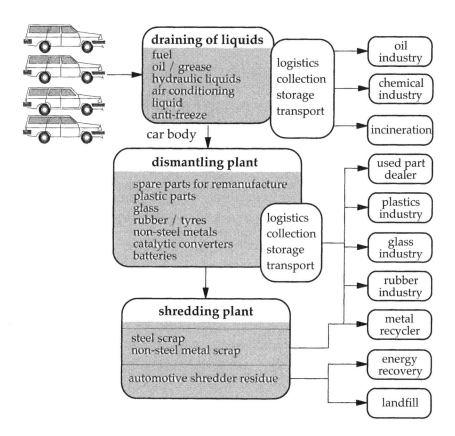

Figure 7.10   ECRIS automotive recycling flow chart.

Plastics is one of the three material groups given priority in the project, the others being rubber and glass. The recycling of steels and metals, accounting for about 75% of the weight of a car, is already well developed; nevertheless, it can be further improved using improved techniques for concentrating and separating metals. During the four year programme, 3000 cars will be dismantled, initially, only Volvos, although in the long run other makes may also be dismantled. Apart from taking in end-of-life cars, the plant is also planned to accept processing waste from Swedish manufacturing plants, and used components disposed of by Volvo owners in special bins at each Volvo dealer will also be accepted.

It must be remembered that the automotive industry is today facing the task of improving recycling rates on vehicles produced 15-20 years ago, which were designed when plastics were considered as throw-away materials. In the future, however, polymer-containing parts will be removed from the hulk before it is shredded, thanks to improved methods of design for recycling being developed in the course of the project. Systems for transport, dismantling and sorting will be developed within the project.

The dismantling process first drains the car of fluids. Then, the car body is brought to the dismantling plant, where material is separated into eight categories, leaving a metal shell that is compressed and fragmented. Recovered plastics are sorted into five categories: ABS, PP, PE, EPDM/PP blends from bumpers, and ABS/PC blends. There is also a significant quantity of open-cell PUR foam that could be recycled into new products such as the undersides of car mats, although at present this material is not recycled due to the complicated construction of parts containing PUR in older cars. Furthermore, there is no capacity to chemically recycle thermosets such as PUR in Sweden. Smaller amounts of PVC are also present in cabling and as surface layers on wood-fibre door panels. These complicated material combinations are highly unlikely in future car design, but still present a problem for current automotive recycling. The ECRIS project will produce valuable information for designers on how to design future car parts to facilitate recycling and increase material recovery.

To be able to operate, each active recycling site will have to be certified as providing consistent quality material for specified applications. Volvo lays no claim on the certifying function. *"This should be a function independent of industry"*, declares Mrs Fräjdin Hellqvist.

Volvo is currently not using much post-consumer material in its cars, for three main reasons: the very limited availability of recycled material, the lack of accurate characterisation methods and the rapid down-cycling of certain plastics that makes them unsuitable for automotive applications. Priorities for environmental improvement wholly favour lightweight design, which in many cases may be in conflict with recyclability. For example, fuel consumption is a sales argument, but so are recyclability and security. These three factors have to be weighed against each other to attain a good design compromise. Automobile manufacturers do not want to trade safety and performance for recyclability or for the use of recycled materials.

The Volvo definition of recycled material is rather stringent and in-plant regrind is not included. Current use of recycled material in Volvo cars is presented in Table 7.1. The strategic goal is to have a significant percentage of post-consumer plastics in cars as soon as the reliability and durability can be guaranteed. Volvo believes it unlikely that post-consumer material in its current state of development will be used to a significant extent in automotive applications. On the other hand, if Volvo were not to use a considerable amount of recycled material in its own products it realises that confidence in the performance of recycled plastics would be damaged, thus discouraging potential users of secondary material.

Table 7.1   Components made from recycled plastics materials in Volvo cars.

Volvo 900 Series		
Component	Material	Weight (kg)
Absorbents	PUR	1
	Total	1

Volvo 800 Series		
Component	Material	Weight (kg)
Luggage compartment panels	PP	5
Absorbents	PUR	1
	Total	6

Volvo 400 Series		
Component	Material	Weight (kg)
Wheel housing	PP	4,1
Various panels	PP	1,5
Heat shield	PET	0,2
	Total	5,7

S40 Series		
Component	Material	Weight (kg)
Front wheel housing	PP	2*0,495
Bumper side support	PP	4*0,083
Luggage compartment flooring	GMT	3,4
Various panels	PP	0,841
Ventilation housing	ABS	4*0,081
	Total	5,887

Life cycle assessment

It is important to recognise that in spite of being cast as a major environmental villain, the car is also one of the most important means of transportation. There is capital knowledge within the company as well as traditions that cannot be overturned from one day to the next. *"For Volvo it is both a matter of optimising its products and processes with respect to the environment within the existing structure and to develop new ideas"*, states Mrs Fräjdin Hellqvist.

As a strategic support tool for decision making, a Life Cycle Analysis method (Environmental Priority Strategies in product design (EPS)) has been developed in a project co-ordinated by the Federation of Swedish Industries called the Product Ecology Project [14]. It aims at developing user-friendly PC-based tools for the calculation of total environmental impact of products "from the cradle to the grave". The system will also serve as a database for environmental information within the company. Volvo is introducing the EPS system as a tool in the chassis and body departments and a number of people have been trained in its use. In the Environmental Concept Car project, the EPS system was used for all calculations.

EPS
Environmental Priority Strategies in product design

Basic Principle

Environmental load index* • Quantity

Environmental Load Value (ELV)
(single score value serving as a base for decisions
on product and production alternatives)

Features:
Life Cycle Inventory
Classification
Characterisation
Valuation

Available Software:
LCA Inventory Tool   (Chalmers Industriteknik)
EPS Enviro-Accounting Method   (Swedish Environmental Research Institute
                                                     and the Federation of Swedish Industries)

Requirements:
• based on global approach (for supplementary regional and local adaptation)
  including the use of natural resources while linked to an economic framework
• possible to simply repeat the calculations
• enables sensitivity and error analysis
• supports validated and motivated judgements about environmental effects

* The unit for the environmental load indices is the ELU (Environmental Load Unit)/kg, referring to
raw materials, substances, elements or semi-manufactured products. Other units can be used, e.g.
ELU/m² and ELU/part

Early supplier involvement

A great deal of material and processing know-how concerning plastics resides with the sub-suppliers. To better profit from their skills, cut lead times in product development and improve material use, Volvo is moving development capacity and responsibility to its sub-suppliers. Increased insight into assembly operations and responsibility for the testing of materials and sub-assemblies gives the sub-suppliers influence over the development process at an earlier stage. They can communicate limitations and possibilities of their technology in order to develop appropriate performance and quality specifications. By integrating the suppliers in the development process, too-stringent requirements leading to less efficient design solutions and unacceptable production losses can be avoided. When possible, Volvo tries to integrate totally sub-suppliers of paramount importance into its own organisation. Early sub-supplier involvement (ESI) is, however, a delicate subject for an OEM. Regardless of the amount of responsibility that is given to a sub-supplier, the OEM is responsible for the performance of the final product. Preserving control over safety issues and exterior design is therefore of major concern to Volvo. Interior or less-visible parts are delegated to the sub-suppliers while issues of safety, quality and exterior design are still controlled by Volvo.

7.2.3 AB KONSTRUKTIONS-BAKELIT

In the context of early supplier involvement and the creation of an efficient recycling network, a company like AB Konstruktions-Bakelit (KB) has a important function to fill: it has considerable materials and processing knowledge for a company of its size. Furthermore it has a long tradition of Design For Manufacturing and Assembly (DFM&A) which can be extended to Design for Disassembly and Recycling. KB is well-equipped, possessing a laboratory for product testing and materials characterisation that gives it a strong advantage compared to its competitors, especially when new materials enter the market.

Reinforced polyamides for automotive applications

The polymer that KB processes in the largest volume is glass-fibre reinforced polyamide 66 (PA 66+GF). Due to stringent requirements from OEMs, certain applications may produce off-specification items. There are three alternatives to handle these parts:

- send them to landfill,
- sell them,
- keep them and develop knowledge and technology for in-plant recycling that will permit good material utilisation.

KB has chosen the latter alternative. The company has accumulated a certain quantity of in-plant recyclate which, rather than selling a relatively expensive plastic too cheaply, it uses it to develop reprocessing, characterisation and secondary applications. A research project was initiated in 1990 with partners at the

- Swiss Federal Institute of Technology in Lausanne(EPFL), Switzerland,
- Royal Institute of Technology (KTH), Stockholm, Sweden,
- DuPont de Nemours, Geneva, Switzerland,

to investigate the possibilities of recycling PA 66+GF [19]. In KB's opinion, skills in recycling will be a major competitive advantage in the future, as pressure on the automotive industry related to environmental regulations concerning plastics and producer responsibility increases. Their product testing and material characterisation laboratory is of great help in this work.

> **KB Recycling Strategy for Automotive Reinforced Polyamide**
>
> **Develop recycling know-how through:**
> - quality control on products and processes
> - performing failure analysis tests
> - developing knowledge about processing and service degradation mechanisms
> - developing design methods for recycled materials
> - developing characterisation methods for durability and reliablity forecasts

Characterisation of recyclate properties

One aim of the project is to develop simple and reliable characterisation methods that enable durability and reliability of recycled material to be predicted. The lack of such methods is one obstacle to the application of recycled material in higher performance applications. Reliable analysis techniques that can be scaled up for industrial application are vital to the acceptance of such materials in industry.

The techniques developed are based on the determination of fibre length and molecular weight on the one hand, and of the thermo-oxidative stability of the matrix material and composite bending strength on the other.

Process-induced fibre shortening and matrix degradation are important phenomena during recycling of reinforced plastics. Processing exposes the material to high shear and temperature fields, resulting in fibre shortening and often in molecular degradation. In the case of PA 66+GF, however, the latter does not always occur; at certain processing temperatures and moisture levels an increase in molecular weight can even be observed, due to post-polymerisation. The decrease in fibre length in recycled items reduces both short-term strength-related properties and long-term creep resistance. The weight average fibre length (l_w) of end-caps based on virgin material, 25% in-plant reground material, and 100% in-plant reground material is shown in Table 7.2 [20].

Table 7.2   Properties of dried samples of PA 66+GF.

SAMPLE	lw (mm)	Mn(g/mole)	σ(MPa)	OPT (°C)
virgin	0.30	17,600	275	317.5
25% in-plant	0.28 *	17,900	259	316.7
100% in-plant	0.24	17,400	258	315.5

* calculated using linear interpolation

The number average molecular weight (Mn) can be used as an indicator of the state of the matrix material. Significant property deterioration in PA 66 is indicated only if Mn falls below approximately 5,000-10,000 g/mole. The Mn values of the recycled samples are considerably higher than this, and so the recyclate does not significantly contribute to deterioration of mechanical properties. Furthermore, the average molecular weight is linked to processing properties such as melt viscosity.

The thermo-mechanical cycle of reprocessing may affect the durability of the matrix material. If heat stabilisers are depleted during repeated processing, the thermal stability of the matrix will be affected and the deterioration of the mechanical properties of recycled items in service may be accelerated. In Table 7.2, the oxidation peak temperature (OPT) measured by differential scanning calorimetry in oxygen is given for the samples studied. The samples containing recyclate exhibit a lower degree of thermal stability, as reflected by lower OPT values. This indicates the consumption of stabilisers during processing. Whereas molecular degradation is negligible, the oxidative stability of recycled PA 66 is reduced; the material requires restabilisation during reprocessing if it is intended for further long-term service.

The mechanical strength (σ) was determined in the direction of flow during moulding (the direction of preferred fibre orientation). The data reported in Table 7.2 indicates, as expected, that recycling results in a reduction in strength. 100% recycling results in an drop of roughly 10%. This can be explained by a lower efficiency of fibre reinforcement due to fibre shortening during reprocessing.

The analytical methods developed within the project are used to map process- and service-induced degradation for the material. Knock-down factors (KD) can be determined for property degradation as a function of process and service parameters.

Figure 7.11 shows the decay of the tensile strength of samples containing 25% and 100% in-plant recycled material against ageing time at 140°C. Fracture toughness and elongation at break were found to be higher than that of virgin material up to an ageing time of 4000 hrs, due to fibre shortening. It is thus the decrease in tensile strength that is the most decisive design criterion within this time range. Material with 25% in-plant recycle remains within design limits for a typical service life for automotive applications (1000 hrs at the indicated temperature).

Information on the effects of ageing on mechanical properties together with the effect of impurities on recyclate properties will allow more efficient tailoring of material to customer needs and also help to avoid over-dimensioned products. Related to the dimensioning of products are the issues of tolerances. One of the prime criteria for acceptance of recycled materials in high performance applications is that dimensional tolerances be within the same range as for virgin material. The results of tests done within the co-operation project indicate that dimensional tolerances remain acceptable, even for 100% in-plant reground radiator end-caps [20].

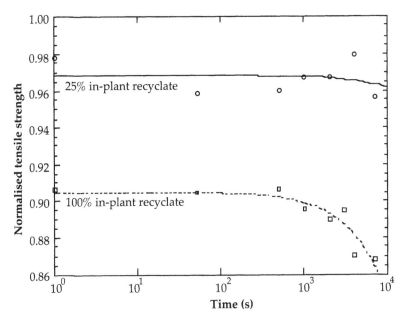

Figure 7.11 Tensile strength for 100% and 25% recycled short-fibre reinforced PA 66 (normalised to the value of virgin material) versus ageing time at 140°C [21].

"The research program is a way for KB to position itself for future changes. It develops competence within the company and gives insight into future possibilities and threats" declares Anders Månson, founder and working chairman of the board.

Sub-supplier involvement

Currently KB works with high-performance compounds in engineering plastics. The major trade volume is in glass-fibre reinforced PA 66. However, the increased demand on recycling may change the future strategy of the company. Anders Edsfeldt, responsible for product development, sees future activities on three levels depending on the quality of the material: high-tech applications for virgin material, medium requirements for in-plant recyclate and low performance applications for post-consumer material.

"Technology and marketing will have to be structured according to these levels, with clients in every field to match the different levels of downgrading of the material", according to Mr Edsfeldt. Anticipating future demand for recycling and reuse of polyamide, KB sees the development of material characterisation methods and applications for reprocessed materials as an excellent platform for future competitivity. Economies of scale would be the key to the success of such operations, and the logistics of off-specification material would be within the capacity of the company.

The automotive industry is currently moving development capacity to their sub-suppliers who are becoming more actively involved in development procedures. For the sub-supplier this implies:

* increased responsibility for development,

* increased liability for function and material,

* responsibility for material and product testing,

* closer and longer-term relations with clients,

* further insight into the operations of the clients,

* freedom to concentrate further on satisfying customer-experienced quality,

* increased need for information on the service conditions of products.

Increased responsibility in product development permits a greater degree of integration of material and processing knowledge into product design. Functional requirements and design can be adapted to better exploit the characteristics of the recycled material, thus avoiding unnecessary production losses due to over-stringent specifications or mould designs poorly adapted to the material characteristics. The material characterisation and testing facilities at KB permits the company to generate unique knowledge which can be used in discussions with material suppliers when selecting materials for applications, as well as to support product development with their clients. This type of communication and co-operation will become of increasingly important as recycling networks are created.

7.2.4 DUPONT DE NEMOURS INTERNATIONAL S.A.

DuPont manages environmental issues in the same manner as traditional safety issues, based on the belief that all accidents and injuries are preventable. The aim of environmental management is to avoid any environmental incidents and to move towards zero waste.

The annual environmental progress report has become an important means of communication with the public and also a tool for further anchoring the company's commitment in-house (Figure 7.12). Putting environmental issues into perspective by means of voluntary reporting is thought to be preferable to risking bad environmental press.

> **The DuPont Commitment**
> **Safety, Health, and the Environment**
>
> • Highest Standards of Performance; Business Excellence
> • Goal of zero injuries, illnesses, and incidents
> • Conservation of energy and natural resources; habitat enhancement
> • Management and employee commitment; accountability
> • Open and public discussion; influence on public policy

Figure 7.12   The DuPont commitment.

DuPont supports various industrial initiatives to improve environmental performance: CEFIC's "Responsible Care®" programme for continuous performance improvement within the fields of health, safety, and environmental protection was implemented at all European sites during 1995. Furthermore, DuPont supports the EMAS regulation and plans to have all major sites registered under this scheme by the end of the 1990's. To follow and influence legislative and regulative changes concerning plastics, DuPont has joined The Association of Plastics Manufacturers in Europe (APME).

The incorporation of environmental activities into daily activities is seen as a key factor in bringing long-term and self-sustaining environmental change into the organisation. Each business area educates, trains, and motivates its staff to understand and comply with applicable laws and the DuPont commitments. Thus the commitment should generate quantifiable results and not only paperwork, which might be the case should a company fail to communicate its commitment down through the hierarchy.

In line with its environmental strategy DuPont has adopted an Integrated Resource Management (IRM) approach to its activities. This includes directives on waste avoidance, waste minimisation, reuse and recycling and the evaluation of environmental decisions and industrial activities with Life Cycle Assessment. The resulting strategy for the DuPont nylon business will be presented, with the emphasis on recycling efforts.

The nylon business

The nylon market can be divided into two segments: resin products and fibre products. The total consumption of nylon in Western Europe was 913,000 tons in 1993, of which 560,000 tons were fibre products divided among textiles, carpets, and industrial fibre [22]. The total annual nylon consumption, as well as that of other engineering plastics, is small compared to that of high-volume commodity thermoplastics; for example, the consumption of LLDPE/LDPE for 1993 was 5,548,000 tons [22]. Engineering plastics therefore constitute a minor share of the total plastics waste. Nevertheless, for an environmentally responsible company it is necessary to develop a clear strategy for the recycling of products and to help their clients do the same.

DuPont advises clients concerning the recycling of in-plant generated production waste within the resin sector and mechanical recycling of post-consumer resin-based products primarily from the automotive sector. In addition, DuPont is creating resin outlets for internally-generated production waste from fibre production and technology for the chemical recycling of post-consumer fibre and resin.

An engineering plastic such as nylon provides properties and a value-margin that make them suitable for recycling. They can either be chemically recycled to give new material with virgin properties, blended with virgin resins to generate material with close-to-virgin properties, or be down-cycled to compete for applications normally occupied by commodity thermoplastics (primarily PP applications). In the latter form the performance-price ratio is in favour of PA (Figure 7.13), given that they are available in equal volumes.

Nylon has a market price of around 5 SFr/kg while that of PP is a little over 1 SFr/kg. The cost of recycling for both materials, given equal availability is 3-4 SFr/kg [23], which suggests that recycled nylon can be sold at a competitive market price compared to the virgin material, while recycled PP becomes three to four times more expensive than its virgin equivalent. The availability of nylon post-consumer waste is, however, much lower than that of PP. This means that nylon recycling logistics will be more energy- and cost-intensive per unit weight than PP, which raises the recycled price to around 7 SFr/kg. Thus an initial advantage is turned into a potential problem for economic recycling.

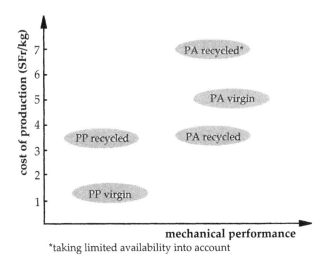

Figure 7.13 Price and performance of virgin and recycled PP and PA (prices calculated as average from Modern Plastics International [24]).

Nylon resins have their primary applications in under-the-bonnet automotive applications and in durable goods, while nylon fibres are primarily used in textiles and carpets. These are all applications of lower volume compared to commodity plastics, with more or less complex material composition, and with disposal patterns that make them difficult to be recovered economically in sufficient amounts.

Mechanical recycling

In co-operation with Kungliga Tekniska Högskolan (KTH, Stockholm, Sweden), the Composite and Polymer Technology Laboratory of the Swiss Federal Institute of Technology (EPFL, Lausanne, Switzerland), and AB Konstruktions-Bakelit, the effects of mixing in-plant production waste with virgin material on short-term properties and on the durability and reliability of short glass-fibre reinforced PA 66 have been studied [19]. Further work will determine the degradation of properties occurring during service of radiator end-caps made from this material. This project was discussed in depth in the previous section on AB Konstruktions-Bakelit.

As previously mentioned, the cost of mechanical reprocessing is not the only factor affecting the final cost of the recycled material. Sufficient supply is a key factor for economic recycling, and collection and separation costs can also be considerable. To examine the economic prerequisites for mechanical recycling of automotive nylon, DuPont is participating in the nylon group of the German PRAVDA initiative (Projekt Autoverwertung Deutsche Automobilindustrie), which is a joint effort of German car producers and plastics suppliers to define the economics of the reuse of automotive plastics. The findings of the first phase of the project, where the only accepted material sources are those within the automotive industry, show that the availability of nylon is a major problem.

Automotive nylon is mainly used in under-the-bonnet applications such as air-intake manifolds and radiator end-caps, and is the largest single market for nylon in Europe [24]. The volume of material collected was not large. Consequently, the automotive nylon stream must be synchronised with supplies from other nylon sources if operations are to be economically viable. Not only must different post-consumer sources be pooled, but the logistics for collection and sorting must be improved to collect a higher share of the waste generated from retired nylon products. The low availability of nylon will also probably imply large transport distances per unit nylon to the operation.

Further areas which need development are materials identification and separation. The batches of nylon studied varied in purity between 20% and 99%. The better the material is sorted at source, the less time- and energy-consuming will be the purification process before reprocessing.

In the second phase of the project all nylon sources were accepted, yet in spite of this it was not possible to collect sufficient amounts for reprocessing tests. Thus the principal lesson from the PRAVDA experience concerning nylon is that insufficient amounts of nylon are available for economic recycling in Europe. In the US, DuPont is already supplying commercial volumes of recycled nylon as mentioned in Chapter 3. A forecast on the evolution of automotive nylon in Europe indicates that sufficient volumes will not be available until 2005 or 2010 (Figure 7.14) [25]. Thus economy in recycling is most likely to be achieved on an international scale, by combining material sources from across national borders.

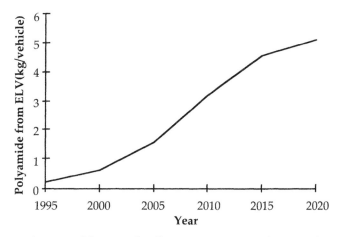

Figure 7.14   A forecast of the growth of heat stabilised glass fibre-reinforced polyamide 66 present in end-of-life vehicles (elv).

Feedstock recycling

Mechanical recycling is not likely to absorb more than 15-20% of the annual total of polymer waste for reasons of availability, markets and cost. In contrast to mechanical recycling, which can be carried out with relatively low capital cost, feedstock recycling requires high capital investment and development costs, and is thus a more suitable activity for a high-tech multinational company able to support high overheads. DuPont has developed a patented chemical recycling process called ammonolysis to recycle PA 6 and PA 66 from post-consumer material as well as from internal plant scrap. The process is depicted in Figure 7.15.

This process targets automotive post-consumer material, discarded carpets with high PA 6 and PA 66 content, and other forms of nylon scrap, in order to generate sufficient volumes for economic recycling. The annual consumption of fibres and textiles is about twice as large as that of resin and would contribute significantly to the recyclable waste volume were it recovered. Today carpets are primarily found in municipal solid waste. While a smaller fraction is collected in kerbside collection schemes [7], most of the carpets are picked up by carpet installers or commercial collectors who send them to landfill.

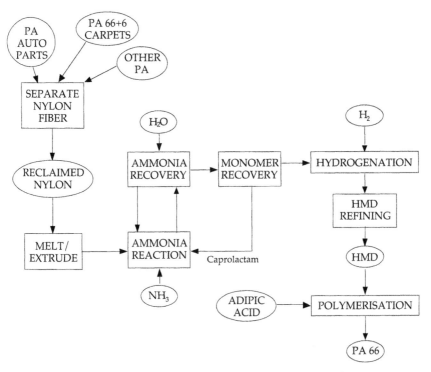

Figure 7.15   The DuPont Ammonolysis process for PA 6 and PA 66 (courtesy DuPont de Nemours).

Since there is already a structure in place to recover carpets it should not be too difficult to redirect this material for recycling. Especially in urban areas where the infrastructure is dense and large volumes of carpets are used in both commercial buildings and in private homes, transport efficiency should be relatively high. DuPont has acted on this opportunity in the USA, where it has initiated a Partnership for Carpet Reclamation (PCR) with carpet mills, carpet dealers, designers and end-users to recover and recycle discarded carpets into new material. Collection networks have been formed in selected regions across North America. Carpets are collected regardless of manufacturer, fibre type, or construction by partnership members and put into collection containers at the carpet dealers. Once a container is full, container collection is co-ordinated between dealers to fill one truckload, which is then taken to a carpet processing centre. At the processing centre, carpets are sorted, identified and evaluated to determine their recycling value. Depending on this value, carpets are recycled into raw material for [26]:

- automotive parts,
- carpet fibre,
- concrete-like products,
- sod-reinforcements,
- wood-replacement products.

For high-quality nylon carpets (which also contain a small proportion of polypropylene) DuPont has developed a density separation process which generates three outputs:

- 98.5% pure nylon,
- 98% pure polypropylene,
- by-product material.

The material generated by this separation process has been approved for direct production of some grades of engineering resins for the manufacture of fans shrouds and other under-the-bonnet automobile parts by the American automotive industry.

The volume of discarded carpets may exceed the capacity of the engineering resin market. In this case, it would be necessary to turn discarded carpets into new carpet fibre. Fibre spinning demands the highest resin quality and tolerances. By chemically recycling the nylon feedstock generated by the separation process, DuPont can obtain a material that as is good as virgin for a recycle content of maximum 50 per cent (ammonolysis only produces hexamethylene diamine (HMD) and the other precursor of PA 66, adipic acid, must be made from oil to obtain PA 66).

The ammonolysis process is scheduled to reach commercial-scale operations in North America before the year 2000. Current recycling cost estimates give a material price that is 10-15 per cent higher than that of a normal PA production plant which would mean a material price of 5.5-5.75 SFr/kg. The total cost of the material will of course depend on the cost and quality of the feedstock. Large quantities of solid waste cannot be purged economically from the process. Fillers or modifiers would have to be removed from the feedstock before recycling to achieve high quality.

Life cycle assessment

DuPont uses life cycle assessment to support decision making and to control the environmental performance of products and processes. In the APME effort to establish life cycle inventory (LCI) modules for different polymers [27] DuPont is taking part in the PA working group. A secrecy agreement is signed with the head expert of an LCA panel consisting of four experts. The head expert then makes process data anonymous and then presents it to the three remaining members of the panel. Once the panel has approved the data, an industrial average is produced, which allows each company to position its product in relation to the mean value of its industry.

A rough estimate of NO_x emissions, energy consumption, and solid waste for short-fibre reinforced PA 66 is shown in Table 7.3. Combined with economic and technological analyses of the feasibility of each method, this information can guide in the choice of a waste disposal alternative.

Table 7.3 Potential environmental impacts of PA 66+GF30 waste disposal alternatives (worst case: all N converted into NO_x).

Waste disposal method	Relative energy use (MJ/kg)	NO_x (%)	Solid waste (%)
Landfill	39	0	100
Combustion	16	15	30
Melt reprocessing	3	0	0
Depolymerisation-repolymerisation	28	0	30

It must be kept in mind that such an estimation is specific to one chosen situation and localisation of production; the result could be different if production were to take place in another country where industrial practices are different.

From an environmental point of view the mechanical route seems to be by far the best alternative, whereas landfill consumes most energy and most landfill capacity.

Furthermore, LCAs are used in evaluations of different product alternatives in collaboration with industrial clients. For example, a life cycle analysis has been performed for Ford, comparing an air-intake manifold in aluminium (manufactured by a particular process at a specified processor) to one made from Zytel® by a specified process and processor [28]. Some of the results are displayed in Figures 7.16 - 7.19 (First published in German Plastics **83**, 3 (1993)).

The results indicates that the nylon alternative has certain advantages from an environmental point of view. Resource consumption in the form of energy and material, global warming gases (CO_2 and N_2O) and consumed landfill volume all speak in favour of the ZYTEL® alternative. Only the contributions to acid rain (NO_x and CO_2) are approximately equal for the two materials.

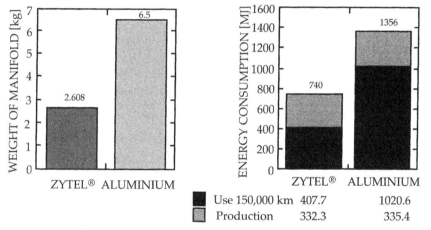

Figure 7.16   Weight of and energy consumption during production and use of air-intake manifolds made from ZYTEL® and aluminium.

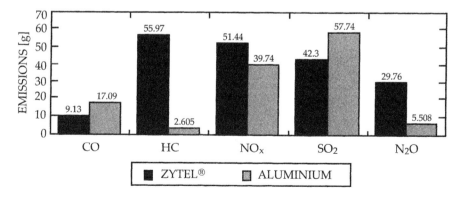

Figure 7.17   Air emissions during production for an air-intake manifold made from ZYTEL® or aluminium.

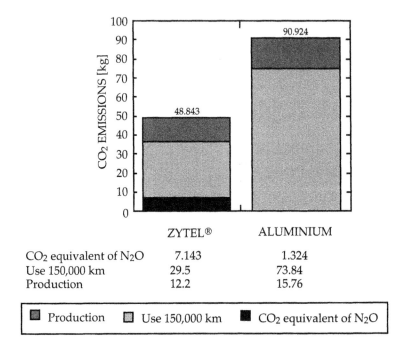

	ZYTEL®	ALUMINIUM
CO_2 equivalent of N_2O	7.143	1.324
Use 150,000 km	29.5	73.84
Production	12.2	15.76

■ Production   ▤ Use 150,000 km   ■ CO_2 equivalent of N_2O

Figure 7.18 Global warming potential for ZYTEL® and aluminium air-intake manifolds during production and use.

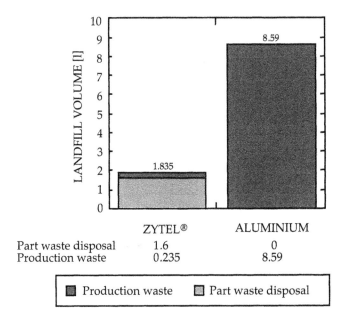

	ZYTEL®	ALUMINIUM
Part waste disposal	1.6	0
Production waste	0.235	8.59

■ Production waste   ▤ Part waste disposal

Figure 7.19 Landfill volume consumed by production waste and final part disposal for an air-intake manifold made from ZYTEL® or aluminium.

7.2.5 FUTURE AUTOMOTIVE RECYCLING NETWORKS

DuPont, AB Konstruktions-Bakelit and their academic partners have investigated the influence of repeated processing and service on the engineering properties of PA 66+GF. The results show that the changes in engineering properties due to mixing 25% of in-plant recycled material with virgin resin are within design limits. This type of material could be used in applications today reserved for 100% virgin material, which would be an important step towards increasing recycling levels in automotive applications. Prior demand for virgin materials mixed with reprocessed has been absent for engineering application, but this practice has been developed into a method of improving the economics of processing without neglecting the final quality of the product. It is thus in the interest of raw material suppliers to assist processing companies in developing upgrading and reprocessing methods for regrind in virgin material.

The remaining question to be solved is the acceptance by OEMs of this kind of material in demanding automotive applications. As mentioned in the section on plastics waste management, feedstock- and mechanically recycled post-consumer nylon has already been approved for automotive under-the-bonnet applications in the US. There is no doubt that when the automotive industry finds quality, supply, and price acceptable, then it will accept this type of material. Due to the presently low availability of post-consumer automotive nylon, the price of a blend of typically 25% recyclate/75% virgin resin would probably be close to that of virgin material. This may seem discouraging, but it must be remembered that this is in an initial stage and automotive design practices and recovery infrastructure is not fully developed. The potential gain of good-will by an OEM such as Volvo in being one of the first automotive producers to apply material with recycle content in higher demand applications may well compensate for the initially high price of high-performance recycled material.

The use of nylons within the automotive sector is rising. More is being used in applications such as air-intake manifolds and other under-the-bonnet components. Sufficient quantities to allow economic recycling within the automotive industry are likely to be available between 2005 and 2010 within the automotive industry, if consumption continues to rise at the current rate [25]. This would explain why greater efforts are being made to chemically or mechanically recycle carpet waste, at least in the US. The experience gained from chemical and mechanical recycling of carpets could be applied to automotive recycling.

In the material chain of the automotive industry two streams of different plastics have been identified: in-plant regrind, or off-specification products, and "post-consumer" materials. The former can be taken care of within the processing community itself. The latter will most likely be the responsibility of the car manufacturer acting through large- or medium-sized compounders and supported with reprocessing know-how from the raw material producers.

There are other potentially more serious environmental issues than automotive recycling: the increasing scarcity of fossil fuels and emission of greenhouse gases. Currently automotive shredder residues constitute 0.4% of landfill waste by weight [22]. The plastics share of this is around 32% [29], accounting for 0.13% of the total annually-generated landfill weight in Western Europe. A 10% increase in the recycling level of the plastics share of ASR would thus only reduce the annual landfill weight by 0.013%. In comparison with the issues of fuel economy and air emissions this might seem to be of minor importance. This does not mean that the subject should be neglected, however: all issues should be addressed simultaneously. The use of plastics in automotive applications is likely to rise given the positive effects their use can have on fuel economy and air emissions. Moreover, the European automotive industry faces take-back legislation and is therefore obliged to work towards recycling.

The commercial opportunities opened up by the incorporation of environmental considerations into management strategies are likely to require resources beyond the capacity of one single company; close networking and the formation of strategic alliances between OEMs and key sub-suppliers as well as the take-back industry is likely to show the highest potential to achieve common environmental and financial goals. The processing industry as well as OEMs need guaranteed quality and supply of recycled materials if they are to be used in higher-demand applications. Raw material producers play an important role in the improvement of the quality of recycled material. With unbeatable knowledge of their materials they can help in the development of separation and cleaning techniques allowing higher quality of recycled material to be achieved. Furthermore, by improving the quality of material they are creating goodwill for their material that is likely to benefit not only themselves, but also the rest of the product chain. Alliances across industrial sectors may also show high potential for the achievement of reasonable waste volumes for recycling.

The activities presented in this case study show the possibilities and difficulties encountered by the plastics-related industries of the automotive sub-supplier chain, as well as by the automotive producers themselves. There is still much to be done, and the discussion on priorities for environmental improvements within transport continues. No doubt, is there great potential for improvements. The automotive and chemical industries are two of the industry sectors exposed to the hardest environmental criticism. They can turn this to their advantage by putting themselves at the forefront of environmental change.

REFERENCES

1. C. Brooling, *Life Cycle Inventory: PP Bags versus Glass Bottles*, Pharmacia & Upjohn Packaging Technology R&D (1994).

2. A. Gregory, Y. Leterrier, Y. Wyser, and J.-A. E. Månson, *Recyclage et Durabilité d'un Film Polymère à Application Pharmaceutique*, Diploma Work, Laboratoire de Technologie des Composites et Polymères, Ecole Polytechnique Fédérale Lausanne, Switzerland (1995).

3. L. A. Utracki, *Polymer Alloys and Blends: Thermodynamics and Rheology*, Carl Hanser Verlag: Munich (1989).

4. G. Haour, F. Szekely, and W. Lee, *How New Requirements on Car Recycling will Jolt the Automotive Industry*, IMD, Lausanne, Switzerland and GE Plastics, Bergen-op-Zoom, The Netherlands, July Draft (1994).

5. S. Nadis, J. J. McKenzie, and L. Ost, *Weighing the Alternatives* in *Car Trouble*, Beacon Press: Boston, USA, p. 72 (1993).

6. M. Heinzl-Rödl, *Future Recycling Opportunities for Automotive Polypropylene in Western Europe*, Laboratoire de Technologie des Composites et Polymères, DMX, Ecole Polytechnique Fédérale de Lausanne (1996).

7. *How to Manage Plastic Waste: Technology and Market Opportunities* , A. L. Bisio and M. Xanthos Eds., Hanser: Munich (1992).

8. A. Arnaud, *Disposal of shredder residue* in proceedings of *R'95*, Geneva, Switzerland, **1**, pp. 35-38 (1995).

9. H. H. Wolf, *Concept of the European Car Manufacturers for the Recycling of Automobiles* in proceedings of *ReC '93*, Geneva, **1**, pp. 1-5 (1993).

10. Renault, *Target: Zero Waste: Automobiles*, The Research & Development Collection, 1.

11. S. Nadis, J. J. McKenzie, and L. Ost, *Car Trouble*, World Resources Institute Guides to the Environment, Beacon Press: Boston, USA (1993).

12. P. Peuch, Association of Plastics Manufacturers in Europe, Avenue E. Van Nieuwenhuyse 4, Box 5, B-1160, Brussels, Belgium: Personal Communication.

13. C. Kendall, *The Treatment of End-of-Life Vehicles - A Partnership Between Industry, Consumers and Public Authorities at the European Level* in proceedings of GLOBEC '96, *9th Global Environment Technology Congress*, Davos, Switzerland, p. 12.1.1 (1996).

14. *The Product Ecology Project: Environmentally-Sound Product Development Based on the EPS System (Environmental Priority Strategies in Product Design)*, Federation of Swedish Industries (1993).

15. O. Giarini and W. R. Stahel, *The Limits to Certainty: Facing Risks in the New Service Economy* , Kluwer Academic Publisher: Amsterdam, The Netherlands, pp. 86-91 (1993).

16. P. Törnqvist, *Producentansvar för bilar*, Miljödepartementet, March 29 (1996).

17. J. Dickinson, *5000 Bilar i återvinningsprojekt*, Naturvårdsverket, June 17 (1996).

18. *Environmentally Compatible Car Scrapping*, Volvo Car Corporation, Göteborg, Sweden, Environmental Report 47 (1994).

19. P.-A. Eriksson, P. Boydell, K. Eriksson, Y. Leterrier, A.-C. Albertson, and J.-A. E. Månson, *Recycling of Automotive Radiator End-Caps* in proceedings of *R'95*, Geneva, Switzerland, **3**, pp. 26-30 (1995).

20. P.-A. Eriksson, A.-C. Albertsson, P. Boydell, K. Eriksson, and J.-A. E. Månson, *Durability of Automotive Radiator End-caps: Effects of Service and Recyclate*, Submitted to Polymer Engineering and Science (1997).

21. P.-A. Eriksson, A.-C. Albertsson, P. Boydell, and J.-A. E. Månson, *Durability of In-plant Recycled Glass-fiber Reinforced Polyamide 66*, Submitted to Polymer Engineering and Science (1997).

22. *Plastics Recovery in Perspective: Plastics Consumption and Recovery in Western Europe 1993*, Association of Plastics Manufacturers in Europe (APME), Brussels, Belgium (1995).

23. P. Boydell, *Revalorisation and Reliability* in proceedings of *Life Cycle Engineering and Recycling of Polymers*, Short Course, EPFL, Lausanne, Switzerland (1995).

24. *Resins Report*, Modern Plastics International, **26**, 1, pp. 61-69 (1996).

25. P.-A. Eriksson, *Mechanical Recycling of Glass Fibre Reinforced Polyamide 66*, PhD Thesis, Departement of Polymer Technology, Royal Institute of Technology, Stockholm, Sweden (1997).

26. P. L. Hauck and R. A. Smith, *Integrated Waste Management* in proceedings of *Dornbirn '95* (1995).

27. I. Boustead, *Eco-Balance Methodology for Commodity Thermoplastics*, Association of Plastics Manufacturers in Europe, Brussels, Belgium, December (1992).

28. M. Schuckert, Th. Dekorsy, I. Pfleiderer, and P. Eyerer, *Developing a Comprehensive Balance of an Automobile Air Intake Manifold*, Kunststoffe - German Plastics, **83**, 3, pp. 16-19 (1993).

29. H.-J. Warnecke, M. Kahrmeyer, and R. Rupprecht, *Automotive Disassembly - A New Strategic and Technologic Approach* in proceedings of *RECY '94*, Erlangen, Germany, pp. 224-233 (1994).

8

AFTERWORD

This book describes the concept of Life Cycle Engineering as it applies to plastics, materials with characteristics that are particularly interesting for applications where weight savings, design freedom and processing cost savings are important, such as packaging and transport.

Plastics consumption is rapidly increasing. Several driving forces lie behind this: economic, environmental and technological. Technology is continuously expanding the properties and applications potential for plastics. This accounts for a steady growth of plastics in transport, packaging and information technology within developed countries. The increased belief in the feasibility of their recycling and their role in source reduction is another driving force that is of importance in these regions. Environment-related advantages of plastics also have positive economic aspects such as the reduction of manufacturing cost, reduced maintenance and reduced transport costs. Thus, even if the stress on environmental issues should decrease in the near future, the findings resulting from environmental considerations have confirmed the value of the approach and thus guaranteed its further use, even for strictly economic or performance reasons.

The burgeoning demand for plastics in threshold countries such as China, India and parts of South America, however, is likely to be a far greater contribution to global plastics consumption within the near future. Thus, even though plastics are likely to contribute to reduced resource consumption in the same way as they do in other regions, the global waste generation of a material that in many cases is considered difficult, if not impossible, to recycle economically is likely to increase dramatically.

The technical feasibility of plastics recovery has nevertheless been proved in several cases. Some products have already shown to constitute economically feasible sources of recycle feedstock. The key technological and infrastructural requirements for economic recycling based on a vital recycling chain, whose criteria of design and construction, collection and sorting, characterisation and reprocessing and application and end-use need to be satisfied in order to ensure economic recycling. Society and industries are moving towards globalisation, as do environmental problems. Plastics are potentially useful in the reduction of resource consumption and environmental impact per useful product or service. There is, however, little chance that efforts of this kind will be able to compensate for increased per capita consumption coupled with the global population increase.

To fully harness the advantages of plastics it is necessary to understand their structure and behaviour during the course of their life cycle in terms of material durability and reliability. Mechanical recycling can generate material with close-to-virgin properties if the purity of the feedstock and upgrading technology is well understood and controlled. These materials can be used in applications where requirements on surface appearance and processing are less stringent. To obtain virgin quality material it is often necessary to convert plastics into their original constituents and then to repolymerise new material by feedstock recycling. If technology and economy does not permit recycling of plastics into new material, the energy content of plastics is high enough to replace fuels in blast-furnaces, cement kilns or to improve the combustion of municipal solid waste.

Life Cycle Assessment can be used to evaluate the environmental viability of each of the recycling technologies. The aim of this environmental tool is to relate the environmental load to the functional unit of a considered product or process by means of an inventory of resources, energy and environmental emissions which is evaluated by an impact assessment. It thus helps to assess technological improvements aimed at ensuring the product or process function with less material, energy and environmental effects. It also helps in evaluating recycling techniques. Life Cycle Assessment can be adapted to industrial conditions in the form of user-friendly design support tools to enable detection and improvement of products and processes already at the concept stage.

It has been found that the integration of environmental considerations into development procedures requires a high level of interdisciplinarity. Integrated product development procedures allow considering environmental issues while respecting traditional development criteria such as time to market, product performance and quality. It is also becoming increasingly common that original equipment manufacturers, in particular automotive, outsource development to sub-suppliers or enter risk-sharing development projects, jointly financed by themselves and the sub-supplier. This permits minimising individual risk and cost as well as fully exploiting the innovative potential of the supplier in the development of new product concepts. For the sub-supplier, this means increased pay-back time and thus higher stress on long-term planning. Design guidelines and design-support software are essential in creating a common language, setting mutually agreed objectives and meeting relevant design criteria.

Material reduction, product life extension, material life extension, process improvements and product management are all best achieved by consulting all relevant parties before the final architecture of a product has been settled.

Increased environmental legislation and arising business opportunities due to environmental concerns have lead to a need for the development of systematised environmental management within companies as well as other sectors of society. Environmental management systems permit the identification of environmental issues relevant to company activities and the establishment of management policies and procedures to improve environmental performance and profitability of companies. Environmental accounting procedures translate environmental issues into costs that can be accounted for in traditional corporate accounting. Environmental reporting is used as a means of communicating environmental progress to the public and contributes to improved corporate image.

A number of plastics-related companies have profited from integrating environmental considerations into their organisation. This integration is reflected in the use of plastics to improve the environmental performance and convenience of products as well as the ensuring of their recycling. Intensified co-operation and openness between the actors within traditional product chains as well as the inclusion of new actors such as dismantlers, reprocessors and users of recycled material is one characteristic of the new structures introduced to meet environmental requirements. To ensure availability of recycled materials, pooling systems have been created, administered in some cases by material suppliers and in others by major material users under particular legislative pressure to do so. Furthermore, several of the industries under environmental pressure have chosen to actively participate in environmental debating in order to affect and guide discussions and to avoid bad press.

Central to all life cycle-oriented strategies is the reduction of emissions related to human activities, energy flows and material consumption. The area of Life Cycle Assessment mainly addresses emissions and energy flows, while Life Cycle Engineering is concerned with material flows in terms of product and material life extension and product reuse and recycling. The implementation of these principles is, however, complex and still not fully understood and technological, environmental, economic or political trade-offs frequently have to be made. There is a clear need for creating a more intimate link between technological and environmental tools such as Life Cycle Engineering and Life Cycle Assessment on the one hand and the dynamics of markets, legislation and public opinion on the other to enable more balanced decisions. The reasons for this are several: people active within these different sectors have different ways of interpreting situations. Consequently they propose different solutions, strategies, technologies and roles for economic sectors. As elucidated by theories on integrated product development, there is a strong need to develop a common language and intense communication between these sectors in order to obtain solutions that take into account fundamental issues such as quality of life and social well-being in the decisions of where and how to apply technological solutions for the better of humanity.

Perhaps one of the more important issues to resolve is the harmonisation of the core values of economic theory and technology. Whereas several areas of industry have already abandoned theories of equilibrium in favour for system dynamics and uncertainty, economic theories still takes equilibrium as the premise for economic analysis. Giarini and Stahel expressed the need for society as a whole to accept uncertainty and risk as legitimate features of life. This would in its turn create the necessary prerequisites for exploring new possibilities and encourage creative solutions to environmental and social problems, which are often regarded more as occasional deviations from the theories of equilibrium.

The last decade has seen the clean-up or closing of several hard-polluting sites due to a strong emphasis on environmental issues among the public and authorities. As these visible environmental problems have been eliminated in several regions, one can not expect that environmental change will have such a strong driving force during the next decade in these regions. What is more likely is a harmonisation between theories, values and interests from different sectors of society into a more homogeneous picture.

On a long-term the stronger confluence of science, technology, economy and social sciences is likely to provide a rich source of intellectual stimulus for the development of new notions of value for society.

The concept of Life Cycle Engineering is not always straightforward. As indicated in this book, implementation requires organisational change as well as the opening up of new communication channels. It is clearly worth the effort, however, judging by the continuously-increasing number of companies adopting life cycle-based strategies as a core management feature for future competitiveness. Rapid technological and legislative changes will surely make the integration of environmental issues a continuous process of learning and renewal for businesses as well as for societies all around the globe.

ACRONYMS

ABS	Acrylonitrile-Butadiene-Styrene Copolymer		EAPS	Environmental Aspects in Product Standards
AP	Acidification		ECA	Aquatic Ecotoxicity
APME	Association of Plastics Manufacturers Europe		ECC	Environmental Concept Car
			ECRIS	Environmental Car Recycling in Scandinavia
ASR	Automotive Shredder Residue		ECT	Terrestrial Ecotoxicity
ASTM	American Society for Testing and Materials		EEC	European Economic Community
BS	British Standard		EL	Environmental Labelling
CAD	Computer Aided Design		ELV	End-of-Life Vehicle
CCE	Competence Centre Environment		EMAS	The EEC Environmental Management and Audit Scheme
CE	Concurrent Engineering			
CEFIC	European Chemical Industry Council		EMS	Environmental Management System
CEN	Comité Européen de Normalisation		EOL	End-of-life
CFC	Chlorofluorocarbon		EPDM	Ethylene Propylene Diene Rubber
COD	Chemical Oxygen Demand: amount of oxidisable material in water		EPE	Environmental Performance Evaluation
DFDA	Design for Disassembly		EPI	Environmental Performance Indicator
DFE	Design for Environment		EPS	Environmental Priority Strategies
DFMA	Design for Manufacturing and Disassembly		ESI	Early Sub-supplier Involvement
DFR	Design for Recycling		GER	Gross Energy Requirement
DG III	Directorate General III		GF	Glassfibre
DIS	Draft International Standard		GIPT	Granulate Injection Paint Technology
DJIA	Dow Jones Industrial Average		GMIA	Good Money Industrial Average
DJUA	Dow Jones Utility Average			
DSC	Differential Scanning Calorimetry		GMT	Glass-Mat Reinforced Thermoplastic
DSD	Duales System Deutchland		GMT	Glass-Mat reinforced Thermoplastic
DTMA	Dynamic Thermomechanical Analysis		GMUA	Good Money Utility Average
EA	Environmental Accounting		GPC	Gel Permeation Chromatography
EA	Environmental Auditing			

GWP	Global Warming Potential	OEM	Original Equipment Manufacturer
HCA	Human Toxicity Air	OIT	Oxidation Induction Temperature
HCS	Human Toxicity Soil		
HCW	Human Toxicity Water	OPT	Oxidation Peak Temperature
HDPE	High Density Polyethylene	OTV	Odour
HIPS	High Impact Polystyrene	PA	Polyamide
HMD	Hexamethylene Diamine	PA 6	polyamide 6
ICC	International Chamber of Commerce	PA 66	polyamide 66
		PC	Polycarbonate
IP	Integrated Processing	PCPW	Post Consumer Plastics Waste
IPD	Integrated Product Development		
IR	Infrared	PCR	Partnership for Carpet Reclamation (DuPont)
ISO	International Standardisation Organisation	PE	Polyethylene
		PERI	Public Environmental Reporting Initiative
IV	Intravenous		
IWM	Integrated Waste Management	PET	Poly(ethene terephthalate)
		PMMA	Poly(methylene methacrylate
LCA	Life Cycle Assessment	POCP	Photochemical Oxidant Formation
LCD	Life Cycle Design		
LCE	Life Cycle Engineering	POM	Poly(oxymethylene)
LCI	Life Cycle Inventory	PP	Polypropylene
LDPE	Low Density Polyethylene	PPE	Poly(phenylene ether)
LLDPE	Linear Low Density Polyethylene	PRAVDA	Projektgruppe Automobilverwertung der Automobilindustrie
MAH	Maleic anhydride		
MNC	Multinational Company	PS	Polystyrene
MRF	Material Recovery Facility	PU	Polyurethanes
MSW	Municipal Solid Waste	PUR	Polyurethane
N_2O	Oxide of Nitrogen: product of denitrifying bacteria, gets into stratosphere	PVC	Poly(vinyl chloride
		PWMI	Polymer Waste Management Institute
NO_X	Oxides of Nitrogen: serious air pollutants in urban areas. Principal source: burning of coal	SAGE	Strategic Advisory Group on the Environment
		SEBS	Styrene-Ethylene-Butylene-Styrene Copolymer
NP	Nutrification		
NR	Natural Rubber	SETAC	Society of Environmental Toxicology and Chemistry
ODP	Oxygen Depletion Potential	SMC	Small and Medium Size Company

SPI	Society of the Plastics Industry
SWB	Solid Waste Burden
TaD	Environmental Terms and Definitions
TC	Technical Committee
TEM	Transmission Electron Microscopy
TGA	Thermo-Gravimetric Analysis
UL	Underwriters Laboratories
UN	United Nations
UNEP	United Nations Environment Programme
UV	Ultraviolet
VCC	Volvo Car Corporation
WBCSD	World Business Council for Sustainable Development
WICE	World Industry Council for the Environment
WLF	Williamson-Landels-Ferry
WWF	World Wide Fund for Nature
ZEV	Zero Emission Vehicle

INDEX

AB Konstruktions-Bakelit 167, 177
   recycling strategy 178
acetolysis 25
acrylonitrile-butadiene-styrene
terpolymer 14
adaptable design 117
additives 18, 55
   antioxidants 19
   blowing agents 21
   environmental aspects of 56
   external light barriers 19
   fillers 21
   flame retardant replacements 57
   hindered amine light stabilisers 20
   internal light barriers 19
   plasticisers 21
   quenchers 19
   radical absorbers 20
   UV absorbers 19
   UV stabilisers 19
adhesion 55
ageing 27, 180
   ageing time shift factor 33
   physical 27
   rate 33
   time-ageing time superposition 33
Agenda 21 2
alcolysis 25
allocation 88
aluminium 189
ammonolysis 186
   material price 188
Arrhenius reaction rate equation 31
Association of Plastics Manufacturers in
Europe 5
auto-heating 26
automotive disassembly 122, 173
automotive shredder residue 165
   share of landfill 192
biodegradation 63
blends 53, 160
Brundtland Commission 2
BS 7750 137
Business Council for Sustainable
Development 136
cars
   disassembly 122
   discarded annually 164
   material composition 165
   new requirements for recycling 168

   plastic grades in 166
   produced annually 164
   product life extension 172
   recycling and waste generation 166
   recycling networks 167
CFC 56
chain scission 26, 55
Chemical Manufacturers' Association 10
closed-loop recycling 88
Club of Rome 1
co-combustion 61
co-production 88
collection 41, 160
   methods 41
combined waste recycling 88
compatibilisation 53
compatible materials 120
concentration 109
concurrent engineering , see integrated
product development
contamination 159
copolymers 14, 54
corporate environmentalism 168
crack growth 35
creep 29, 33
degradation 25, 27
   ASTM standard 63
   factors 23
   matrix 178
   solvolysis 25
   thermo-oxidative 25, 31
density methods 43
depolymerisation 25, 57, 187, 188
design
   adaptable 116
   for assembly and disassembly 105, 116,
   123
   for maintainability 116
   for recycling 105, 121
   material reduction through 112
design tools
   computer-based 106
   guidelines 123
designing with uncertainty 22
disassembly 122, 123, 170
Dow Jones Industrial and Utility
Averages 148
DuPont de Nemours 168
durability 22, 179, 184
   effects of internal stresses 26

materials versus products 115, 121
prediction 22
reprocessing 179
early supplier involvement 181
eco-labelling 79
economics
    disassembly 122
    environmental management 133, 147
    recycling 183
    sorting 46
EEC Environmental Management and
Auditing Scheme 139
EMS, see Environmental Management
System
end-of-life vehicles 164
    future evolution 165
energy recovery 8
Environmental Car Recycling in
Scandinavia 172
environmental management
    certification 139, 140
    economics 133, 147
    legal consequences 140
    loopholes 140
    preparatory review 138
    system elements 134, 138, 139, 143
environmental performance 132
Environmental Priority Strategies in
product design 176
environmental reporting
    core components 143
    guidelines 143-147
    small and medium-sized enterprises
    150
environmental strategy 102, 155
epoxies 14
ethical investments 133
European Chemical Industry Council 143
external load stresses 29
fatigue 34
feedstock recycling 57, 186
fibre shortening 26, 178
flow-induced degradation 26
fracture toughness 35
glass transition 33
global warming potential 92
goal definition and scoping 80
Good Money Industrial and Utility
Averages 148
granulated injection paint technology 111
green Funds 133, 148
Greenpeace 2
gross energy requirement 92
heat of combustion 61
hot runner systems 111

hydrolysis 25
I.V. solutions 156
ICC Business Charter for Sustainable
Development 134
identification 42
incineration 60
integrated product development 104
integrated waste management 48
internal stresses 29
inventory analysis 78
IR spectroscopy 43
knock-down factors 180
legislation 70
    automotive 166, 167
    packaging waste 67, 145
    producer responsibility 131
    take-back regulations 131
life cycle assessment 78, 120, 176, 188
    air-intake manifolds 188-190
    allocation 88
    applications by frequency of use 79
    capital goods 87
    critical surface-time 92
    data collection 88
    data quality 84
    data relevance 97
    databases 97
    definition 78
    delineation 85
    depth of the study 82
    discrepancies 96
    driving forces 79
    environmental profile 92
    equal functionality 83
    equivalency factors 92
    evaluation 94
    external 80
    functional boundary 81
    functional unit 82, 97
    impact assessment 78, 91
    impact categories 91
    impact scores 92, 93
    improvement analysis 78, 95
    internal 80
    inventory 85
    open-loop recycling 86
    process tree 85, 97
    product system 85
    production modules 87
    production unit 83, 156
    qualitative aspects 91
    regional boundary 81
    reliability 95
    scope 81

sensitivity analysis 87
SETAC Code of Practice 85, 91
software 97
standard classification model 91
temporal boundary 81
users 79
validity 95
life cycle cost 103
life cycle engineering 8, 101, 130
    goal 8
    implementation 10
    in product development 102
    methodology 3
    of radiator end-caps 130
life cycle inventory 78, 156
life extension 115
long-term strategic planning 97
long-term viscoelastic response 33
low-temperature oxidation 26
maintainability 116
maleic anhydride 54
master curve 33
material
    selection 107
    separation 120
    substitution 119
    value 102
material compliance 33
material reduction
    in processing 111
    integrated functions 112
    material and process integration 113-114
    miniaturisation 112
    optimised wall thickness 112
    simplified design 112
matrix degradation 178
mechanical recycling 178, 184
    flow scheme 50
    in-plant 48, 184
    melt filtration 50
    multiple-processing degradation 48
    post-consumer 46, 49, 66
    techniques 48, 51
molecular markers 45
multilayer film 158
multiple output process 88
multiple processes 88
municipal solid waste 40
open-loop recycling 86, 88
outdoor weathering 27
oxidation 24
ozone attack 27
ozone depletion potential 92

packaging 155
    recycling 160
    waste 145
partitioning, see allocation
Partnership for Carpet Reclamation 187
performance value 7, 8
pharmaceutical 155
Pharmacia & Upjohn
    attitudes towards plastics 162
photo-oxidation 27
physical ageing 27
plastics
    advantages 5, 164
    categories 6
    consumption 5
    definition of 18
    disadvantages 6
    domestic waste by plastic type 42
    energy content 61
    environmentally degradable 63
    growth 4
    waste 40, 64
Plastics Waste Management Institute 97
polyamide 58, 161
    6, 161
    66 177
    availability 185
    consumption 183
    glass fibre-reinforced 177, 188
    growth in end-of-life vehicles 185
    life cycle assessment 188
    primary applications 184
    recycling economics 183
polycarbonate 15
polyester ether 161
polyethylene 13, 58, 162
polyethylene terephthalate 16, 55, 162
polymer blends 53, 160
polymers
    amorphous 15
    basic classes 14
    basic properties 15
    characterisation methods 18
    degradation 23
    glass transition temperature 17
    life cycle of 24
    oxidation 24
    poisson's ratio 18
    rubbers 14
    semi-crystalline 16
    solvolysis 25
    stiffness 17
    strength 17
    structure 13-14, 16

thermal degradation 25
thermal transitions 16
thermoplastics 13
thermosets 14
viscoelasticity 17
polypropylene 13, 16, 54, 58, 161, 162
polystyrene 13, 15, 58
polyurethanes 14
polyvinyl chloride 13, 58, 161, 162
prediction
    creep 32
    degradation 30
    durability 32
    fatigue 34
    viscoelastic response 34
    WLF 33
pro-active 136
process improvement 103
product development 104
Product Ecology Project 176
product obsolescence 115
Projekt Autoverwertung Deutsche
Automobilindustrie 184
Public Environmental Reporting Initiative
144
PVC, see polyvinyl chloride
pyrolysis 25
qualitative multicriteria analysis 95
quantitative multicriteria analysis 95
recyclability 101
recycling 145, 157, 160, 172
    automotive 68
    characterisation 160, 178
    collection 159
    contamination 159, 160
    criteria for efficiency 157
    economic driving forces 65
    factors for market acceptance 65-68,
    180, 191
    food packaging 69
    IT equipment and consumer
    electronics 68
    knock-down factors 179
    legislation 66
    long-term mechanical performance 161
    networking 192
    problems 52
    radiator end-cap 164
    reprocessing 160, 179
    secondary applications 159
    transportation 160
    wood replacements and interior
    design 69
refilling 110, 162
reliability 22, 184

remanufacturing 117
reprocessing 179
restabilisation 55
reuse 118
Rio environmental summit 2
risk assessment 79
selective dissolution 44
SETAC, see Society of Environmental
Chemistry and Toxicology
shift factor 33-34
Society of Environmental Chemistry and
Toxicology 80
Society of the Plastics Industry 42
solid waste burden 92
solvent absorption 27
solvolysis 25, 58
sorting 41, 159
    economics 46
    macro sorting 41
    sourcing problems 46
specific properties 108
stabilisers 20
stress corrosion 25
stress relaxation 29
structural recovery 27
styrene-butadiene copolymer 14
styrene-butadiene rubber 14
styrene-ethylene-butylene-styrene 158
substance flow analysis, 79
superposition
    time-ageing time 33
    time-temperature 32
sustainable development 2
thermal analysis 161
thermo-oxidative degradation
    accelerated testing 32
    activation energy 31
    Arrhenius reaction rate equation 30
    long-term prediction 31
thermolysis 59
time-temperature superposition 33
transportation 110, 163
unsaturated polyesters 14
value analysis 104
Verein Deutscher Ingenieure 122
vertical boundary 81
vis-breaking 56
voluntary plastic container coding system
42
Volvo Car Corporation 167
    communication 169
    Competence Centre Environment 169
    early supplier involvement 177
    Environmental Car Recycling in
    Scandinavia 170

  Environmental Concept Car 171
  life cycle assessment 175
  material selection 171
  product development 170
  Product Ecology Project 171
  use of recycled plastics 175
  Volvo Environmental Management
  System 169
waste 3, 145, 155, 163
  management 64
  packaging 163
  produced in Western Europe 3
  recovery 40
weight reduction 108
  examples 110
  knock-on effects 110
  packaging 109
Williams-Landel-Ferry 33
World Business Council for Sustainable
Development 2, 136
World Industry Council for the
Environment 136

Printed and bound by CPI Group (UK) Ltd, Croydon, CR0 4YY

03/10/2024

01040414-0020